CAMBRIDGE LIBRARY COLLECTION

Books of enduring scholarly value

Physical Sciences

From ancient times, humans have tried to understand the workings of the world around them. The roots of modern physical science go back to the very earliest mechanical devices such as levers and rollers, the mixing of paints and dyes, and the importance of the heavenly bodies in early religious observance and navigation. The physical sciences as we know them today began to emerge as independent academic subjects during the early modern period, in the work of Newton and other 'natural philosophers', and numerous sub-disciplines developed during the centuries that followed. This part of the Cambridge Library Collection is devoted to landmark publications in this area which will be of interest to historians of science concerned with individual scientists, particular discoveries, and advances in scientific method, or with the establishment and development of scientific institutions around the world.

Memoir and Correspondence of Caroline Herschel

Memoir and Correspondence of Caroline Herschel (1876) contains the letters and diaries of the celebrated astronomer Caroline Herschel (1750–1848), edited by her great-niece, Mary Herschel. Caroline was born in Hanover to a musician father and an illiterate mother who did not want her daughter to be educated. However Caroline's brother William, an organist employed in Bath, persuaded their mother to allow Caroline to join him there. She left for England in 1772 to live with William, to whom she remained devoted all of her life. In Bath, William turned towards telescope-making and astronomy, to such effect that in 1781 he discovered the planet Uranus. He was appointed 'the King's astronomer' in 1782, and Caroline, trained by William, continued to work at his side as a scientist in her own right. Between them, they discovered eight comets and raised the number of recorded nebulae from a hundred to 2,500.

Cambridge University Press has long been a pioneer in the reissuing of out-of-print titles from its own backlist, producing digital reprints of books that are still sought after by scholars and students but could not be reprinted economically using traditional technology. The Cambridge Library Collection extends this activity to a wider range of books which are still of importance to researchers and professionals, either for the source material they contain, or as landmarks in the history of their academic discipline.

Drawing from the world-renowned collections in the Cambridge University Library, and guided by the advice of experts in each subject area, Cambridge University Press is using state-of-the-art scanning machines in its own Printing House to capture the content of each book selected for inclusion. The files are processed to give a consistently clear, crisp image, and the books finished to the high quality standard for which the Press is recognised around the world. The latest print-on-demand technology ensures that the books will remain available indefinitely, and that orders for single or multiple copies can quickly be supplied.

The Cambridge Library Collection will bring back to life books of enduring scholarly value (including out-of-copyright works originally issued by other publishers) across a wide range of disciplines in the humanities and social sciences and in science and technology.

Memoir and Correspondence of Caroline Herschel

EDITED BY MARY HERSCHEL

CAMBRIDGE
UNIVERSITY PRESS

CAMBRIDGE UNIVERSITY PRESS

Cambridge, New York, Melbourne, Madrid, Cape Town, Singapore,
São Paolo, Delhi, Dubai, Tokyo, Mexico City

Published in the United States of America by Cambridge University Press, New York

www.cambridge.org
Information on this title: www.cambridge.org/9781108013666

© in this compilation Cambridge University Press 2010

This edition first published 1876
This digitally printed version 2010

ISBN 978-1-108-01366-6 Paperback

CAROLINE HERSCHEL

Caroline Herschel.

ÆTAT 92.

MEMOIR AND CORRESPONDENCE

OF

CAROLINE HERSCHEL.

BY

MRS. JOHN HERSCHEL.

WITH PORTRAITS.

LONDON:

JOHN MURRAY, ALBEMARLE STREET.

1876.

INTRODUCTION.

FAMILIAR to all as is the name this volume bears, it is not without hesitation that the following pages are given to the world. To subject the memorials of a deeply earnest life to the eyes of a generation over-crowded with books, raises a certain amount of diffidence.

Of Caroline Herschel herself most people will plead ignorance without feeling ashamed, and yet may we not assert that Caroline Herschel is well worth knowing.

Great men and great causes have always some helper of whom the outside world knows but little. There always is, and always has been, some human being in whose life their roots have been nourished. Sometimes these helpers have been men, sometimes they have been women, who have given themselves to help and to strengthen those called upon to be leaders and workers, inspiring them with courage, keeping faith in their own idea alive, in days of darkness,

When all the world seems adverse to desert.

These helpers and sustainers, men or women, have all the same quality in common—absolute devotion and unwavering faith in the individual or in the cause. Seeking nothing for themselves, thinking nothing of themselves, they have all an intense power of sympathy, a noble love of giving themselves for the service of others, which enables them to transfuse the force of their own personality into the object to which they dedicate their powers.

Of this noble company of unknown helpers Caroline Herschel was one.

She stood beside her brother, William Herschel, sharing his labours, helping his life. In the days when he gave up a lucrative career that he might devote himself to astronomy, it was owing to her thrift and care that he was not harassed by the rankling vexations of money matters. She had been his helper and assistant in the days when he was a leading musician ; she became his helper and assistant when he gave himself up to astronomy. By sheer force of will and devoted affection, she learned enough of mathematics and of methods of calculation, which to those unlearned seem mysteries, to be able to commit to writing the results of his researches. She became his assistant in the workshop ; she helped him to grind and polish his mirrors ; she stood beside his telescope in the nights of midwinter, to write down his observations, when the very ink was frozen in the bottle. She kept him alive by her care ; thinking nothing of herself, she lived for him. She loved him, and believed in him, and helped him, with all her heart and with all her strength. She might have become a distinguished woman on her own account, for with the "seven-foot Newtonian sweeper" given to her by her brother, she discovered eight comets first and last. But the pleasure of seeking and finding for herself was scarcely tasted. She "minded the heavens" for her brother; she worked for him, not for herself, and the unconscious self-denial with which she gave up her own pleasure in the use of her "sweeper," is not the least beautiful feature in her life. She must have been witty and amusing, to judge from her books of "Recollections." When past eighty, she wrote what she called "a little history of my life from 1772—1778 " for her nephew, Sir John Herschel, the son of her brother William, that he might know something of his excellent grandparents, as well as of the immense difficulties which his father had to surmount in his life and labours. It was not to tell about herself, but of others, that she wrote them. There is not any good biography of Sir William Herschel, and the incidental revelations of him in these Recollections are valuable. They show how well he deserved the love and devotion she rendered to him. Great as were his achievements in science, and his genius, they were borne up and ennobled by the beauty and worth of his own inner life.

These memorials of his father and his aunt were

much valued by Sir John Herschel, and they are carefully preserved by the family along with her letters. The perusal of them is like reading of another world. The glimpses of the life of a soldier's family in Hanover at the time the Seven Years' War was going on are very touching. Both father and mother must have been remarkable persons, and the sterling quality of character developed in William and Caroline Herschel was evidently derived from them. All the family seem to have been endowed with something like touches of genius, but William and Caroline were the only two who had the strong back-bone of perseverance and high principle which made genius in them fulfil its perfect work.

Her own recollections go back to the Great Earthquake at Lisbon; she lived through the American War, the old French Revolution, the rise and fall of Napoleon, and all manner of lesser events and wars. She saw all the improvements and inventions, from the lumbering post waggon in which she made her first journey from Hanover, to the railroads and electric telegraphs which have intersected all Europe, for she lived well down into the reign of Victoria. But her work of " minding the heavens " with her brother engrossed all her thoughts, and she scarcely mentions any public event.

Her own astronomical labours were remarkable, and in her later life she met with honour and recognition

from learned men and learned societies ; but her
dominant idea was always the same—" I am nothing,
I have done nothing ; all I am, all I know, I owe
to my brother. I am only the tool which he shaped
to his use—a well-trained puppy-dog would have
done as much." Every word said in her own praise
seemed to be so much taken away from the honour
due to her brother. She had lived so many years in
companionship with a truly great man, and in the
presence of the unfathomable depths of the starry
heavens, that praise of herself seemed childish
exaggeration.

The Letters and Recollections contained in this
volume will show what she really was. She would
have been very angry if she could have foreseen their
publication, yet, in consideration of the great interest
they possess, we hope to be justified for making
known to the world such an example of self-sacrifice
and perseverance under difficulties.

The spelling has been modernised,—an old lady
who had discovered eight comets might be allowed
to spell in her own way ; but it is pleasanter to read
what is written in an accustomed manner. A word
has been altered occasionally where the sense required
it, otherwise no change has been made, and as little
has been added as was possible, and only with the
view of giving a slight connecting thread of narrative.

If these Recollections convey as much pleasure

to the readers of them as they have given to the Editor, they will feel that they have gained another friend in Caroline Lucretia Herschel.

December, 1875.

<hr>

NOTE.

WHEN past ninety a second memoir was undertaken, and in order to encourage her to continue it her niece, Lady Herschel, wrote to her as follows:— .... "Now, my dearest aunt, you must let me make an earnest petition to you, and that is, that you will *go on* with your memoir until you leave England and take up your residence in Hanover. How can I tell you how much my heart is set upon the accomplishment of this work? .... You know you cannot be *idle* while you live. But indeed, if I could tell you the influence which a short account by a stranger of *your* labours with your dear Brother had upon me when a child, and of my choosing *you* (then so unknown to me) as my guiding star and example, you would understand how the possession of such a record by your own hand would make me almost believe in auguries and presentiments, and perhaps inspire some future generations more worthily, as the record would be more genuine."

August 9, 1841.

May we not echo this hope, and feel indeed that "SHE BEING DEAD YET SPEAKETH."

M. C. H.

CONTENTS.

CHAPTER I.

PAGE

EARLY LIFE IN HANOVER—MUSICAL TALENTS OF HER BROTHER WIL-
LIAM—MARRIAGE OF HER SISTER—THE REGIMENT ORDERED TO
ENGLAND—HER FATHER'S INDUSTRY—TYPHUS FEVER—CONFIRMA-
TION—DEATH OF FATHER—ACCOMPANIES WILLIAM TO ENGLAND . 1

CHAPTER II.

LIFE IN BATH—HEIMWEHE—THE MIGHTY TELESCOPE—LAST PERFORM-
ANCE IN PUBLIC—CASTING THE GREAT MIRROR—WILLIAM HERSCHEL
GOES TO LONDON — MADE ROYAL ASTRONOMER — REMOVAL TO
DATCHET — ACCIDENTS — GRANT OF £2,000 — LIFE AT SLOUGH —
LETTERS FROM HANOVER—DISCOVERY OF A COMET 29

CHAPTER III.

WILLIAM HERSCHEL'S MARRIAGE—DISCOVERY OF THE EIGHTH COMET—
EXTRACTS FROM DAY-BOOK AND DIARY—VISIT TO BATH—RETURN
TO SLOUGH—RESIDES AT UPTON—ILLNESS—FEAR OF BLINDNESS . 78

CHAPTER IV.

EXTRACTS FROM DIARY — WILLIAM HERSCHEL KNIGHTED — FAILING
HEALTH — HER BROTHER'S PORTRAIT — DEATH OF ALEXANDER—
DEATH OF SIR WILLIAM — HER RETURN TO HANOVER — RECOL-
LECTIONS WRITTEN AT HANOVER 118

CHAPTER V.

RETROSPECTION—LIFE IN HANOVER—HER HUMILITY—HER WORKS—
MADE HON. MEMBER OF THE ROYAL ASTRONOMICAL SOCIETY

PAGE

—HER SWEEPINGS—BLANK IN HER LIFE AT HANOVER—LETTERS
TO LADY HERSCHEL—LETTERS BETWEEN HER AND HER NEPHEW
—VISIT FROM HER NEPHEW—FINISHES HER CATALOGUE OF THE
NEBULÆ 141

CHAPTER VI.

LIFE IN HANOVER CONTINUED — LETTERS BETWEEN HER AND HER
NEPHEW—HER WILL—FIRST CHAPTER OF HER HISTORY—RECEIVES
THE GOLD MEDAL OF THE ROYAL ASTRONOMICAL SOCIETY—FEAR-
FUL STORM—HER PORTRAIT—HER NEPHEW'S MARRIAGE—PREPA-
RATION FOR HER DEATH—PAGANINI—HER NEPHEW KNIGHTED—
LADY HERSCHEL'S DEATH—RETROSPECTION 196

CHAPTER VII.

LETTERS FROM THE CAPE—HON. MEMBER OF THE ROYAL ASTRONOMICAL
SOCIETY—CATALOGUE OF OMITTED STARS—LETTERS—SATURN AND
HIS SIXTH SATELLITE — HER NEPHEW'S VISIT—HON. MEMBER OF
THE ROYAL IRISH ACADEMY—EXTRACTS FROM DAY-BOOK—ANEC-
DOTE OF THE OLD TELESCOPE — CHRISTMAS IN HANOVER—GOLD
MEDAL FROM THE KING OF PRUSSIA — DECLINING STRENGTH —
DEATH—FUNERAL 262

TO BINDER.

Portrait of Caroline Herschel. *Frontispiece.*
Portrait of Sir William Herschel, after the original by Abbott, in
the National Portrait Gallery, to face p. 118.
Herschel's Forty-foot Telescope, to face p. 29.

CAROLINE HERSCHEL.

CHAPTER I.

CAROLINE LUCRETIA HERSCHEL was born at Hanover on the 16th of March, 1750. She was the eighth child and fourth daughter of Isaac Herschel, by Anna Ilse Moritzen, to whom he was married in August, 1732. The family consisted of ten children, four of whom died in early childhood.

A memorandum in the handwriting of Isaac Herschel, transcribed by his daughter in the original German at the beginning of her Recollections, traces the family back to the early part of the seventeenth century, about which time, it appears that three brothers *Herschel* left Moravia on account of their religion (which was Protestant), and became possessors of land in Saxony. One of these brothers, Hans, was a brewer at Pirna, a little town two miles from Dresden, and the father of two sons, one of whom, Abraham by name, was born in 1651, was the father of the abovementioned Isaac, and the grandfather of Caroline

B

Lucretia Herschel. Abraham Herschel was employed in
the royal gardens at Dresden, he received commissions
from various quarters on account of his taste and skill
as a landscape gardener. Of his four children, Euse-
bius, the eldest, appears to have kept up little or no intercourse with his family after the father's death in 1718.
The second child, Apollonia, married a landed proprietor, Herr von Thümer. Benjamin, the second son, died
in his third year ; and Isaac, the youngest, was born
14th of January, 1707, and was thus an orphan at
the early age of eleven years. His parents wished
him to be a gardener like his father, but a passionate
love of music led him to take every opportunity of
practising on the violin, besides studying music under
a hautboy-player in the royal band. When he was
about one and twenty he resolved to seek his fortune,
and went to Berlin, where the style of hautboy playing was so little to his taste that he soon left it, and
went to Potsdam, where he studied for a year under
the celebrated Cappell Meister Pabrïch, the means
for so doing being supplied by his mother and sister ;
his brother, as he quaintly remarks, contenting himself with writing him letters in praise of the virtue
of economy ! In July, 1731, he went to Brunswick,
and in August to Hanover, where he at once obtained
an engagement as hautboy-player in the band of the
Guards, and in the August following he married as
above stated.

The family group to which Miss Herschel's autobiography introduces us consisted of—

1. Sophia Elizabeth, born in 1733. [Afterwards Mrs. Griesbach.]
2. Henry Anton Jacob, born 20th November, 1734.
(4) 3. Frederic William, born 15th November, 1738.
(6) 4. John Alexander, born 13th November, 1745.
(8) 5. Carolina Lucretia, born 16th March, 1750; and
(10) 6. The little Dietrich, born 13th September, 1755.

With the exception of frequent absences from home which attendance on a regiment made inevitable, the family life went on smoothly enough for some years, the father taking every opportunity, when at home, to cultivate the musical talents of his sons, who depended for the ordinary routine of education on the garrison school, to which all the children went from the age of two to fourteen. Here the splendid talents of *William* early displayed themselves, and the master confessed that the pupil had soon got beyond his teacher. Although four years younger than Jacob, when the two brothers had lessons in French, the younger had mastered the language in half the time needed by the elder, and he in some measure satisfied his eager desire for knowledge by attending out of school hours to learn all that his master could teach of Latin and arithmetic. At fourteen he was an excellent performer both on the oboe and violin.

The first serious calamity recorded was the irreparable injury caused to the father's health by the hardships of war. After the battle of Dettingen (June 16th, 1743) the troops remained all night on the field, which was soaked by heavy rains. The unfortunate bandmaster lay in a wet furrow, which caused a complete loss of the use of his limbs for some time, and left him with an impaired constitution and an asthmatical affection which afflicted him to the end of his life. During the dark times of the Seven Years' War, the little Caroline, then her mother's sole companion, often heard this grievous trouble spoken of, and the shadow of it cast a gloom over her childish recollections, most of which are of a sombre character. At three years old she was a deeply interested participator in all the family concerns, and of that period she writes :—

"It must have been in 1753 when my brother [Jacob, aged 19] was chosen organist to the new organ in the garrison church ; for I remember my mother taking me with her the first Sunday on its opening, and that before she had time to shut the pew door, I took fright at the beginning of a preludium with a full accompaniment, so that I flew out of church and home again. I also remember to

have seen my brother William confirmed in his new oböi-
sten uniform."

The next interesting event was the marriage of the
eldest daughter, who was living with a family at
Brunswick, and whom her sister says she had never seen
until she came home to be married. The bridegroom,
Mr. Griesbach, also a musician in the Guard, found no
favour in the eyes of his sister-in-law, and it is evidently some satisfaction to her to have been told that
her father never cordially approved the match,

"for . . . he knew him at least to be but a very middling
musician, and this alone would have been enough for my
father's disapprobation."

Great preparations were made for

" providing and furnishing a habitation (which happened to
be in the same house where my parents lived), which they
did in as handsome a manner as their straitened income
would allow, and to which my dear brothers took delight in
contributing to the best of their ability. I remember how
delighted I was when they were showing me the pretty
framed pictures with which my brother William had decorated
his sister's room, and heard my mother relate afterwards,
that the brothers had taken two months' pay in advance for
the wedding entertainment. . . Though for stocking a family
with household linen my mother was prepared at all times,
as perhaps never a more diligent spinner was heard of; but
to keep pace with the wishes of my dear brothers, by whom
my sister was, as well as by her parents, exceedingly
beloved—the whole family were kept for a time in an

agreeable bustle to see that nothing that could give either pleasure or comfort might be wanting in her future establishment. . . . The fête (without which it would have been scandalous in those days to get married) ended with a ball, at which I remember to have been dancing among the rest without a partner."

A little later, when war troubles broke up the household, and the bride returned to her mother, we are told :

" my sister was not of a very patient temper, and could not be reconciled to have children about her, and I was mostly, when not in school, sent with Alexander to play on the walls or with the neighbour's children, in which I seldom could join, and often stood freezing on shore to see my brother skating on the Stadtgraben (town ditch) till he chose to go home. In short, there was no one who cared anything about me."

The earthquake which destroyed Lisbon on the 1st November 1755, was strongly felt at Hanover, and became closely associated in the poor little girl's mind with the trials and troubles which shortly afterwards fell upon the family. She says :—

" One morning early I was with my father and mother alone in the room, the latter putting my clothes on, when all at once I saw both standing aghast and speechless before me ; at the same time my brothers, my sister, and Griesbach came running in, all being panic-struck by the earthquake."

For a little while the family enjoyed a peaceful interval, during which the extraordinary proficiency of

his two eldest sons was a growing source of delight
to the father, whose utmost ambition was to see them
become accomplished musicians; while the wider
flights of William met with his most cordial sympathy. The following passage is one of the very
few which reflect the brighter side of the picture :—

" My brothers were often introduced as solo performers
and assistants in the orchestra of the court, and I remember that I was frequently prevented from going to
sleep by the lively criticism on music on coming from a
concert, or conversations on philosophical subjects which
lasted frequently till morning, in which my father was a
lively partaker and assistant of my brother William by
contriving self-made instruments. . . . . Often I would
keep myself awake that I might listen to their animating
remarks, for it made *me so happy* to see *them so happy.* But
generally their conversation would branch out on philosophical subjects, when my brother William and my father
often argued with such warmth, that my mother's in-
terference became necessary, when the names Leibnitz,
Newton, and Euler sounded rather too loud for the repose
of her little ones, who ought to be in school by seven in
the morning. But it seems that on the brothers retiring
to their own room, where they shared the same bed, my
brother William had still a great deal to say; and frequently
it happened that when he stopped for an assent or reply, he
found his hearer was gone to sleep, and I suppose it was
not till then that he bethought himself to do the same.
    " The recollection of these happy scenes confirms me in
the belief, that had my brother William not then been interrupted in his philosophical pursuits, we should have had
much earlier proofs of his inventive genius.   My father

was a great admirer of astronomy, and had some knowledge
of that science; for I remember his taking me, on a clear
frosty night, into the street, to make me acquainted with
several of the most beautiful constellations, after we had
been gazing at a comet which was then visible. And I
well remember with what delight he used to assist my
brother William in his various contrivances in the pursuit
of his philosophical studies, among which was a neatly
turned 4-inch globe, upon which the equator and ecliptic
were engraved by my brother."

Towards the end of the year 1755 the regiment
was under orders for England, and the little household was at once broken up. A place in the court
orchestra had been promised to Jacob, but the vacancy did not, unfortunately, occur in time, and
he was obliged to smother his discontent, lower his
ambition, and accept a place in the band with his
younger brother. At length the sad hour of parting
arrived :—

"In our room all was mute but in hurried action; my
dear father was thin and pale, and my brother William
almost equally so, for he was of a delicate constitution and
just then growing very fast. Of my brother Jacob I only
remember his starting difficulties at everything that was
done for him, as my father was busy to see that they were
equipped with the necessaries for a march. . . . The
whole town was in motion with drums beating to march:
the troops hallooed and roared in the streets, the drums
beat louder, Griesbach came to join my father and brothers,
and in a moment they all were gone. My sister fled to

her own room.   Alexander went with many others to
follow their relatives for some miles to take a last look.   I
found myself now with my mother alone in a room all in
confusion, in one corner of which my little brother Dietrich
lay in his cradle; my tears flowed like my mother's, but
neither of us could speak.   I snatched a large handkerchief
of my father's from a chair and took a stool to place it at
my mother's feet, on which I sat down, and put into her
hands one corner of the handkerchief, reserving the opposite
one for myself; this little action actually drew a momen-
tary smile into her face . . .   My father left half his
pay for our support in the hands of an agent in Hanover,
but Griesbach, instead of following my father's example,
gave up his lodging and brought his wife with her goods
and chattels to her mother, which arrangement was no
small addition to our uncomfortable situation."

Even at this early age, it is not difficult to trace
in these childish recollections the influence of that
intense affection for her brother William which
made him more and more the centre of all her
interests; next to him, her father filled a large
place in her heart. Of the long year of separation,
nothing is recorded. At last Jacob arrived (having
" out of aggravation" got permission to resign his
place when the hoped-for vacancy in the orchestra
had been otherwise filled) he had travelled by post,
while his father and brother, " who never forsook
him for self-consideration," were still toiling wearily
on the march home.

" My mother being very busy preparing dinner, had

suffered me to go all alone to the parade to meet my
father, but I could not find him anywhere, nor anybody
whom I knew; so at last, when nearly frozen to death, I
came home and found them all at table. My dear brother
William threw down his knife and fork, and ran to
welcome and crouched down to me, which made me
forget all my grievances. The rest were so happy . . at
seeing one another again, that my absence had never
been perceived."

The visit to England appears to have further developed the love of show and luxury which painfully
distinguished Jacob, who must needs import specimens of English goods and English tailoring, while
all that William brought back was a copy of Locke
on the Human Understanding, the purchase of which
absorbed all his private means, as he never willingly
asked his father for a single penny. But it was becoming apparent that he had not the physical strength
to continue in the Guard during war time, and after
the disastrous campaign of 1757, and the defeat at
Hastenbeck,* 26th July, 1757 (between 20 and 30
miles from Hanover), his parents resolved to remove
him—a step apparently attended by no small difficulty, as our faithful chronicler narrates :—

"I can now comprehend the reason why we little ones
were continually sent out of the way, and why I had only by
chance a passing glimpse of my brother as I was sitting at

---

* The Duke of Cumberland's army suffered severely in this battle.

the entrance of our street-door, when he glided like a
shadow along, wrapped in a great coat, followed by my mother
with a parcel containing his accoutrements. After he had
succeeded in passing unnoticed beyond the last sentinel
at Herrenhausen he changed his dress. . . . My brother's
keeping himself so carefully from all notice was undoubtedly
to avoid the danger of being pressed, for all unengaged
young men were forced into the service. Even the clergy,
unless they had livings, were not exempted."

During these times of public and private peril, the
little girl was sent regularly to the garrison school with
her brother Alexander till three in the afternoon, when
she went to another school till six, to learn knitting.

" From that time forward I was fully employed in providing my brothers with stockings, and remember that the
first pair for Alexander touched the floor when I stood
upright finishing the front. Besides this my pen was
frequently in requisition for writing not only my mother's
letters to my father, but for many a poor soldier's wife in
our neighbourhood to her husband in the camp : for it
ought to be remembered that in the beginning of the last
century very few women, when they left country schools,
had been taught to write."

In addition to these occupations, she was called upon
to make herself useful when the fastidious Jacob
honoured the humble table with his presence, " and
poor I got many a whipping for being awkward at supplying the place of footman or waiter." The sight of
her mother constantly in tears ; the prolonged absence

of her father; the sister's unhappiness at being home-
less when about to become a mother; all these circumstances combined to sadden the personal recollections
of a time of almost unsurpassed national calamity.
After the loss of the battle at Hastenbeck, the Recollections thus conclude this period.

"Nothing but distressing reports came from our army,
and we were almost immediately in the power of the
French troops,* each house being crammed with men.
In that in which we were obliged to bewail in silence
our cruel fate, no less than 16 privates were quartered,
besides some officers who occupied the best apartments, and
this lasted for about two years [a note of later date says
"not so long"] before the town was liberated."

A gap occurs here, between the years 1757 and
1760, several pages having been torn out in both the
original "Recollections" and the unfinished memoir
commenced in 1840. In the former, a sentence beginning "the next time I saw him [Jacob] was when
he came running to my mother with a letter, the

* "While the King of Prussia was warring in the south of Germany, an
army of 60,000 Frenchmen under Marshal d'Estrées was directed upon
Hanover, and occupied in the first place the Prussian dominions lying upon
the Rhine. . . . . . d'Estrées had been to a certain degree successful in an
action at Hastenbeck, on the Weser, and had forced Cumberland to retreat.
That commander continued to yield ground incessantly, leaving Hanover and
Magdeburg unprotected. . . . . He concluded with Richelieu the convention
of Closter Severn, by which he engaged that . . . . the Hanoverian troops
should continue inactive in their quarters near Stade. Hostilities were to be
suspended, and no stipulation was made respecting the Electorate of Hanover.
That country was accordingly plundered without mercy, and subjected to
enormous contributions."—*Annals of France, Encyclopædia Metro olitana.*

contents of which," remains unfinished, and the narrative recommences with : " After reading over many
pages, I thought it best to destroy them, and merely
to write down what I remember to have passed in our
family." Accordingly there is no record of anything
preserved during this interval until May, 1760, when
the head of the family returned to it for good—broken
in health and worn out by hardships to which he was
no longer equal, but strong in purpose and devoting
himself at once to the musical education of his children and to giving lessons to the numerous pupils who
soon came to seek instruction from so excellent a
master. Jacob returned for the second time from
England at the end of 1759, and obtained the place
of first violin in the court orchestra. As usual the
appearance of this member of the family caused a
general upset of domestic comfort, for

" when he came to dine with us, it generally happened that
before he departed his mother was as much out of humour
with him as he was at the beefsteaks being hard, and
because I did not know how to clean knives and forks with
brickdust."

The younger children made great progress under
their father's careful training, and with all her propensity for seeing the dark side, the daughter's
recollections of this period afford glimpses of a
tolerably happy household. If it was " a helpless
and distracted family " to which, as she writes, her

father returned, those epithets could ill apply to the
father himself, for there is abundant evidence that he
was a man of no ordinary character—one who, in spite
of constant suffering of a most distressing kind, persisted in hard work to the very end, and who set his
children a noble example of patience, unselfishness,
and self-denial. To the last, as his daughter records,—

" Copying music employed every vacant moment, even
sometimes throughout half the night, and the pen was not
suffered to rest even when smoking a pipe, which habit he
indulged in rather on account of his asthmatical constitution than as a luxury ; for, without all exception, he was the
most abstemious liver I ever have known ; and in every
instance, even in the article of clothing, the utmost frugality
was observed, and yet he never was seen otherwise than
very neat. . . . With my brother [Dietrich] now a
little engaging creature of between four and five years old—
he was very much pleased, and [on the first evening of his
arrival at home] before he went to rest, the Adempken
(a little violin) was taken from the lumbering shelf and
newly strung and the daily lessons immediately commenced.
 . . . I do not recollect that he ever desired any other
society than what he had opportunities of enjoying in many
of the parties where he was introduced by his profession ;
though far from being of a morose disposition ; he would
frequently encourage my mother in keeping up a social
intercourse among a few acquaintances, whilst his afternoon
hours generally were taken up in giving lessons to some
scholars at home, who gladly saved him the troublesome
exertion of walking. . . . He also found great pleasure
in seeing Dietrich's improvement, who, young as he was,
and of the most lively temper imaginable, was always ready

to receive his lessons, leaving his little companions (with whom our neighbourhood abounded) with the greatest cheerfulness to go to his father, who was so pleased with his performances that—I think it must have been in October or November—he made him play a solo on the Adempken in Rake's concert, being placed on a table before a crowded company, for which he was very much applauded and caressed, particularly by an English lady, who put a gold coin in his little pocket.

" It was not long before my father had as many scholars as he could find time to attend, for some of those he had left behind returned to him again, and several families who had sons of about the age of my little brother, became his pupils and proved in time very good performers. And when they assembled at my father's to make little concerts, I was frequently called to join the second violin in an overture, for my father found pleasure in giving me some-times a lesson before the instruments were laid by after practising with Dietrich, for I never was missing at those hours, sitting in a corner with my knitting and listening all the while."

A serious interruption of this and all other occupa-tions was caused by a severe attack of typhus fever which in the summer of 1761 threatened to be fatal, and

" reduced my strength to that degree that for several months after I was obliged to mount the stairs on my hands and feet like an infant; but here I will remark that from that time to this present day (June 5, 1821) I do not remember ever to have spent a whole day in bed."

In spite of her strong objections to learning, the

worthy mother had too correct a view of her duties
to stand in the way of the necessary preparation for
her daughter's confirmation, who was accordingly, but
not without complaints at the loss of time, released
from her household avocations for this purpose.
Alexander, who had been taken as a sort of apprentice by Griesbach, was now of an age to turn his
great musical talents to profitable account, and returned to Hanover, where he obtained the somewhat
mysterious situation of Stadtmusicus (Town Musician),
the duties of which office involved

" little else to do but to give a daily lesson to an apprentice
and to blow a Corale from the Markt Thurm; so that nearly
all his time could be given to practice and receiving instruction from his father. There was no doubt but that he
would soon become a good violin player, for his natural
genius was such that nothing could spoil it."

Although the absent brother William kept up
regular correspondence with Hanover, many of his
letters were written in English and addressed to
Jacob, on such subjects as the Theory of Music, in
which the family in general could not participate.
Year after year went by, and William showed no
inclination to leave England, to which country he
was becoming more and more attached; the poor
father, who felt his strength steadily declining, became
painfully eager for his return. On the 2nd April,
1764, they were thrown into "a tumult of joy" by his

appearance among them. The visit was a very
brief one, offering no hope of any intention to settle
in Hanover; the father was well aware that he
at least could not look forward to another meeting
on earth, while to the poor little unnoticed girl, this
visit and its attendant circumstances stood out in
her memory as fraught with anguish, which even
her unskilled pen succeeds in representing as a grief
almost too deep for words.

" Of the joys and pleasures which all felt at this long-
wished-for meeting with my—let me say my *dearest*
brother, but a small portion could fall to my share; for
with my constant attendance at church and school, besides
the time I was employed in doing the drudgery of the
scullery, it was but seldom I could make one in the group
when the family were assembled together.

" In the first week some of the orchestra were invited
to a concert, at which some of my brother William's compositions, overtures, &c., and some of my eldest brother
Jacob's were performed, to the great delight of my dear
father, who hoped and expected that they would be turned
to some profit by publishing them, but there was no printer
who bid high enough.

" Sunday the 8th was the—to me—eventful day of my
confirmation, and I left home not a little proud and encouraged by my dear brother William's approbation of my
appearance in my new gown."

Not only was she disappointed in her fervent hope
that the longed-for brother would not come at the
very time when she was obliged to be much from

home, but several of the precious days of his stay were spent in a visit to the Griesbachs at Coppenbrügge, and the Sunday fixed for his departure was the very day on which she was to receive her first communion.

" The church was crowded and the door open : the Hamburger, Postwagen passed at eleven, bearing away my dear brother, from whom I had been obliged to part at 8 o'clock. It was within a dozen yards from the open door ; the postilion giving a smettering blast on his horn. Its effect on my shattered nerves, I will not attempt to describe, nor what I felt for days and weeks after. I wish it were possible to say what I wish to say, without feeling anew that feverish wretchedness which accompanied my walk in the afternoon with some of my school companions, in my black silk dress and bouquet of artificial flowers—the same which had served my sister on her bridal day. I could think of nothing but that on my return I should find nobody but my disconsolate father and mother,.for Alexander's engagements allowed him to be with us only at certain hours, and Jacob was seldom at home except to dress and take his meals."

From the state of hopeless lethargy in which the poor sister describes herself as going mechanically about her daily tasks after that memorable day, she was roused by a calamity which affected all alike. The father had a paralytic seizure the August following, by which he lost the use of his right side almost entirely, and although he so far recovered as to be able still to receive pupils in his own house, he never

regained his former skill on the violin, and was reduced to a sad state of suffering and infirmity; a few
months later he was pronounced to be in a confirmed
dropsy. Changes of abode, not always for the better;
anxieties, on account of Alexander's prospects and
Jacob's vagaries; disappointment, at seeing his
daughter grow up without the education he had
hoped to give her; were the circumstances under
which the worn-out sufferer struggled through the
last three years of his life, copying music at every
spare moment, assisting at a Concert only a few
weeks before his death, and giving lessons until he
was obliged to keep wholly to his bed. He was released from his sufferings at the comparatively early
age of sixty-one on the 22nd March, 1767, leaving
to his children little more than the heritage of his
good example, unblemished character, and those
musical talents which he had so carefully educated,
and by which he probably hoped the more gifted of
his sons would attain to eminence.

Miss Herschel describes herself as having fallen into
"a kind of stupefaction," which lasted for many
weeks after the loss of her father, and the awakening
to life had little of hope in the present or promise for
the future, so far as she could see then. At the age
of seventeen she had learned little beyond the first
elements of education, and she was now deprived of
the one friend who encouraged and sympathised with

her desire for better instruction. The parents had never agreed on the subject. "When I had left school," she writes,

"My father wished to give me something like a polished education, but my mother was particularly determined that it should be a rough, but at the same time a useful one; and nothing farther she thought was necessary but to send me two or three months to a sempstress to be taught to make household linen. Having added this accomplishment to my former ingenuities, I never afterwards could find leisure for thinking of anything but to contrive and make for the family in all imaginable forms whatever was wanting, and thus I learned to make bags and sword-knots long before I knew how to make caps and furbelows. . . . . . . My mother would not consent to my being taught French, and my brother Dietrich was even denied a dancing-master, because she would not permit my learning along with him, though the entrance had been paid for us both; so all my father could do for me was to indulge me (and please himself) sometimes with a short lesson on the violin, when my mother was either in good humour or out of the way. Though I have often felt myself exceedingly at a loss for the want of those few accomplishments of which I was thus, by an erroneous though well-meant opinion of my mother, deprived, I could not help thinking but that she had cause for wishing me not to know more than was necessary for being useful in the family; for it was her certain belief that my brother William would have returned to his country, and my eldest brother not have looked so high, if they had had a little less learning.

    *   *   *   *   *   *

But sometimes I found it scarcely possible to get through with the work required, and felt very unhappy that

no time at all was left for improving myself in music or
fancy-work, in which I had an opportunity of receiving
some instruction from an ingenious young woman whose
parents lived in the same house with us. But the time
wanted for spending a few hours together could only be
obtained by our meeting at daybreak, because by the time
of the family's rising at seven, I was obliged to be at my
daily business. But during the summer months of 1766
very few mornings passed without our spending a few hours
together, to which I was called by my friend's loud cough
at her window by way of notice that she was ready for me
[she could not sleep, and was glad of my company. I lost
her soon after, for she died of consumption]. Though
I had neither time nor means for producing anything immediately either for show or use, I was content with keeping samples of all possible patterns in needlework, beads,
bugles, horsehair, &c., for I could not help feeling troubled
sometimes about my future destiny; yet I could not bear
the idea of being turned into an Abigail or housemaid, and
thought that with the above and such like acquirements
with a little notion of Music, I might obtain a place as
governess in some family where the want of a knowledge of French would be no objection."

It was with the same object of fitting herself to
earn her bread that, after her father's death, she
obtained permission to go for a month or two to learn
millinery and dress-making; her eldest brother Jacob,
before leaving them to join William at Bath, having
graciously given his consent, " if it was only meant
to learn to make my own things, but positively for-
bidding it for any other purpose." The following

account of this episode shows how customary such apprenticeship was among young ladies of good family, as a part of their education :—

"My mother found some difficulty in persuading the lady to whom I wished to go, to receive me without paying the usual premium, but at last she gave me leave to come on paying one thaler per month. I felt myself rather humbled on going the first time among twenty-one young people with an elegant woman, Madame Küster, at their head, directing them in various works of finery. Among the group were several young ladies of genteel families, and as I came there on rather reduced terms, I expected that I should be kept in the back ground, doing nothing but the plain work of the business ; but contrary to my fears, I gained in the school-mistress a valuable friend . . . Here I found myself daily happy for a few hours and one of the young women,* after a lapse of thirty-five years, when I was introduced to her at the Queen's Lodge, received me as an old acquaintance, though I could but just remember having sometimes exchanged a nod and smile with a sweet little girl about ten or eleven years old. But I soon was sensible of having found what hitherto I had looked for in vain—a sincere and disinterested friend to whom I might have applied for counsel and comfort in my deserted situation."

A proposal from Jacob that Dietrich, whom the father on his deathbed had specially commended to his care, should be sent to England, caused his mother the utmost distress, on account of his being still too young to be confirmed ; but her scruples were overcome and

---

* Afterwards Madame Beckedorff, Miss Herschel's most valued friend in after years.

Dietrich was despatched in the summer as soon as a fitting escort could be found.

" But what was yet more aggravating was, that the loss of his company was supplied by a country cousin whom my mother permitted to spend the summer with us in order to have the advantage of my mother's advice in making preparation for her marriage. . . . This young woman, full of good-nature and ignorance, grew unfortunately so fond of me that she was for ever at my side, and by that means I lost what little interval of leisure I might then have had for reading, practising the violin, &c., entirely. Besides this, I was extremely discomposed at seeing Alexander associating with young men who led him into all manner of expensive pleasures which involved him in debts for the hire of horses and carioles, &c., and I was (though he knew my inability of helping him) made a partaker in his fears that these scrapes should come to the knowledge of our mother.

" My time was, however, filled up pretty well with making household linen, &c., against Jacob's return. . . .

. . It was not, however, till the middle of the following summer that we saw him again, and I suppose his stay must have been prolonged on account of waiting till he had had the honour of playing before their Majesties, for which (in consequence of having composed and dedicated a set of six sonatas to the Queen) he was informed he would receive a summons. . . . After this his salary was augmented by 100 thalers," and the promise of not being overlooked in future.

[Note.—Before I leave this subject I cannot help remembering the sacrifices these good people were making to pride. They played nowhere for money, for even when in 1768 (I think it was) the King's theatre was first opened *to the Public,* and the Court orchestra was called upon to play there, they did it without any emolument, so that there was no way left to increase their small salaries but by giving a few subscription concerts in the winter, or by teaching. So much, by way of apology, for the emigration of part of my family to England.]

"We passed the winter in the utmost quiet, except when Alexander took it into his head to entertain gentlemen in his own apartment, which always made my mother very cross, else in general nothing disturbed us in our occupation. My mother spun, I was at work on a set of ruffles of Dresden-work for my brother Jacob, whilst Alexander often sat by us and amused us and himself with making all sorts of things in pasteboard, or contriving how to make a twelve-hour Cuckoo clock go a week. . . . As my mother saw that Dietrich's confirmation was still uncertain, she insisted on having him back again. . . . Accordingly at the end of July they [Jacob and Dietrich] arrived, and Dietrich entered school again immediately," but remained only until his confirmation the following Easter.

A new direction was suddenly given to all their plans by the arrival of letters from the absent brother William, who proposed that his sister should join him at Bath—

. . . "to make the trial if by his instruction I might not become a useful singer for his winter concerts and oratorios, he advised my brother Jacob to give me some lessons by way of beginning; but that if after a trial of two years we should not find it answer our expectation he would bring me back again. This at first seemed to be agreeable to all parties, but by the time I had set my heart upon this change in my situation, Jacob began to turn the whole scheme into ridicule, and, of course, he never heard the sound of my voice except in speaking, and yet I was left in the harassing uncertainty whether I was to go or not. I resolved at last to prepare, as far as lay in my power, for both cases, by taking, in the first place, every opportunity when all were from home to imitate, with a gag between my teeth, the solo parts of concertos, *shake and all*, such as I had heard them play on the violin; in consequence I had gained a tolerable execution before I knew how to sing. I next began to knit ruffles, which were intended for my brother William in case I remained at home—else they were to be Jacob's. For my mother and brother D. I knitted as many cotton stockings as would last two years at least."

Jacob remained with his family until the following July, when he returned to Bath, this time taking Alexander with him for two years' leave of absence, the young Dietrich being deemed competent not only to supply his place in the orchestra, but also to attend his private pupils.

Nothing is recorded in the interval between Jacob's return to Hanover in the autumn and the long expected arrival of William in April, 1772, except one of the changes of abode, which were of such frequent occurrence, involving abundance of employment in making and altering articles of household use, which afforded some relief to the conscientious daughter, who was sorely troubled by uncertainty as to her duty in the matter of going to England or staying with her mother, although the latter had given her consent to the change.

"In this manner" [making prospective clothes for them] "I tried to still the compunction I felt at leaving relatives who, I feared, would lose some of their comforts by my desertion, and nothing but the belief of returning to them full of knowledge and accomplishments could have supported me in the parting moment, which was much embittered by the absence of my brother Jacob, who was with the Court which attended on the Queen of Denmark at the *Görde*, where my brother Dietrich had also been for some time, and but just returned when my brother William, for whose safety we had for several weeks been under no small apprehension, at last quite unexpectedly arrived. . . . His stay at Hanover could at the utmost not be prolonged above a fortnight. . . . My mother had consented to my going with him, and the anguish at my leaving her was somewhat alleviated by my brother settling a small annuity on her, by which she would be enabled to keep an attendant to supply my place." They all went over to Coppenbrügge "to see my sister—I to take leave of her; the remaining time was wasted in an unsatisfactory correspondence: the letters from my brother Jacob expressed nothing but regret and impatience at being thus disappointed, and, without being able to effect a meeting, I was obliged to go without receiving the consent of my eldest brother to my going. . . . .

        *   *   *   *   *

"But I will not attempt to describe my feelings when the parting moment arrived, and I left my dear mother and most dear Dietrich on Sunday, August 16th, 1772, at the Posthouse, and after travelling for six days and nights on an open (in those days very inconvenient) Postwagen, we were on the following Saturday conveyed in a small open vessel from the quay at Helvotsluis on a stormy sea, to the packet boat, which lay two miles distant at anchor; from which we were again obliged to go in an open boat to be set ashore, or rather thrown like balls by two English sailors, on the coast of Yarmouth.* For the vessel was almost a wreck, without a main and another of its masts.

" After having crawled to one of a row of neat low houses, we found the party previously arrived from the ship devouring their breakfast; several clean-dressed women employed in cutting bread and butter (from fine wheaten loaves) as fast as ever they could. One of them went upstairs with me to help me to put on my clothes, and after taking some tea we mounted some sort of a cart to bring us to the next place where diligences going to London would pass. But we had hardly gone a quarter of an English mile when the horse, which was not used to go in what they called the shafts, ran away with us, overturning the cart with trunk and passengers. My brother, another person, and myself all throwing themselves out, I flying into a dry ditch. We all came off however, with only the fright, owing to the assistance of a gentleman who, with his servant, was accompanying us on horseback. These persons had come in the packet with us, and it was settled not to part till in London, where we arrived at noon on the 26th at an inn in the City. Here we remained till the evening of the 27th. My brother having business at the West-end of the town, left me under the care of our fellow travellers; but after his return, in the evening when the shops were lighted up, we went to see all that was to be seen in that part of London, of which I only remember the opticians' shops, for I do not think we stopped at any other.

* The other version calls it " from Helvot to Harrige " = Harwich.

" The next day the mistress of the inn lent me a hat of her daughter's—mine was blown into one of the canals of Holland, for we had storms by land as well as at sea—and we went to see St. Paul's, the Bank, &c., &c. Mem : only the outside, except of St. Paul's and the Bank, and we were never off our legs, except at meals in our inn. Towards evening we went to the West of the town, where, after having called on Despatch Secretary Wiese and his lady (Mr. Wiese conducted our correspondence with Hanover) we went to the inn, from whence we at ten o'clock in the evening started by the night coach for Bath on the 28th of August. . . . . . After taking some tea I went immediately to bed, and I did not awake till the next day in the afternoon, when I found my brother had but just left his room. I for my part was, from the privation of sleep for eleven or twelve days (not having above twice been in what they called a bed) almost annihilated."

END OF RECOLLECTIONS, VOL. I.

The only allusion to this journey in Sir W. Herschel's Journal is the brief entry :—" August 16, 1772. Set off on my return to England in company with my sister."

SIR WILLIAM HERSCHEL'S FORTY-FOOT TELESCOPE AT SLOUGH.

[To face page 29.

CHAPTER II.

AT the time when William Herschel brought his
sister back with him to Bath, he had established himself there as a teacher of music, numbering among his
pupils many ladies of rank.   He was also organist of
the Octagon Chapel, and frequently composed anthems,
chants, and whole services for the choir under his
management.   On   the   retirement   of   Mr.   Linley
(father of the celebrated singer, afterwards the beautiful Mrs. Sheridan) from the direction of the Public
Concerts, he at once added this to his other avocations, and was consequently immersed in business of
the most laborious and harassing kind during the
whole of the Bath season.   But he considered all this
professional work only as the means to an end ; devotion
to music produced income and a certain degree of leisure,
and these were becoming every day more imperatively
necessary.   Every spare moment of the day, and
many hours stolen from the night, had long been

devoted to the studies which were compelling. him to become himself an observer of the heavens. Insufficient mechanical means roused his inventive genius ; and, as all the world knows, the mirror for the mighty forty-foot telescope was the crowning result. To his pupils he was known as not a music-master alone. Some ladies had lessons in astronomy from him, and, at the invitation of his friend Dr. Watson, he became a member of a philosophical society then recently started in Bath, to which he for several years contributed a great number of papers on various scientific subjects. It soon came to pass that the gentlemen who sought interviews with him, asking for a peep through the wonderful tube, carried stories of what they had seen to London, and these were not long in finding their way to St. James's.

It was thus at the very turning-point of her brother's career that Caroline Herschel became his companion and fellow-worker. No contrast could be sharper than that presented by the narrow domestic routine she had left to the life of ceaseless and inexhaustible activity into which she was plunged ;— unless, indeed, it be that presented by the nature of the events she has to record, and the tone in which they are recorded. For ten years she persevered at Bath, singing when she was told to sing, copying when she was told to copy, " lending a hand " in the workshop, and taking her full share in all the stirring

and exciting changes by which the musician became the King's astronomer and a celebrity; but
she never, by a single word, betrays how these
wonderful events affected her; nor ever indulges
in the slightest approach to an original sentiment,
comment, or reflection not strictly connected with
the present fact. Whether it be to record the
presentation of the "golden medal," or the dis-
honesty of the incorrigible Betties who then, and
till her life's end, so sorely tried her peace of
mind, there is no difference in the style or spirit
of the " Recollections." Partly as apology and
partly as complaint, the one grievance is harped
on, even when fifty years' experience might have
convinced her that she had done something more
for herself and the world than earn her bread by
her own labour. " In short," she writes, " I have
been throughout annoyed and hindered in my endeavours at perfecting myself in any branch of
knowledge by which I could hope to gain a credit-
able livelihood." It is seldom, however, that she is
diverted from the main theme to write about herself
otherwise than incidentally, and in a note addressed
to her nephew, she says:—" My only reason for
saying so much of myself is to show with what
miserable assistance your father made shift to obtaining the means of exploring the heavens."

" On the afternoon of August 28th, 1772, I arrived
with my brother at his house No. 7, New King Street,
Bath, where we were received only by Mr. Bulman's family,
who occupied the parlour floor, and had the management of
his servant and household affairs. My brother had formerly
boarded with them at Leeds, whence, on Mr. Bulman's
failure in business, they had removed to Bath, where my
brother procured for him the place of Clerk at the Octagon
Chapel. . . . On our journey he had taken every opportunity to make me hope to find in Mrs. Bulman a well-
informed and well-meaning friend, and in her daughter, a
few years younger than myself, an agreeable companion.
But as I knew no more English than the few words which I
had on our journey learned to repeat like a parrot, it may
be easily supposed that it would require some time before I
could feel comfortable among strangers. But as the season
for the arrival of visitors to the Baths does not begin
till October, my brother had leisure to try my capacity for
becoming a useful singer for his concerts and oratorios,
and being very well satisfied with my voice, I had two or
three lessons every day, and the hours which were not
spent at the harpsichord were employed in putting me in
the way of managing the family. . . . On the second
morning, on meeting my brother at breakfast, he began
immediately to give me a lesson in English and arithmetic,
and showed me the way of booking and keeping accounts
of cash received and laid out. . . . By way of relaxation we talked of astronomy and the bright constellations with which I had made acquaintance during the
fine nights we spent on the Postwagen travelling through
Holland.

" My brother Alexander, who had been some time in
England, boarded and lodged with his elder brother, and

with myself, occupied the attic. The first floor, which was
furnished in the newest and most handsome style, my
brother kept for himself. The front room containing the
harpsichord was always in order to receive his musical friends
and scholars at little private concerts or rehearsals. . . . .
Sundays I received a sum for the weekly expenses, of which
my housekeeping book (written in English) showed the
amount laid out, and my purse the remaining cash. One of
the principal things required was to market, and about six
weeks after coming to England I was sent alone among
fishwomen, butchers, basket-women, &c., and I brought
home whatever in my fright I could pick up. . . . My
brother Alex, who was now returned from his summer engagement, used to watch me at a distance, unknown to me,
till he saw me safe on my way home. But all attempts to
introduce any order in our little household proved vain,
owing to the servant my brother then had—a hot-headed
old Welshwoman. All the articles, tea-things, &c., which
I was to take in charge, were almost all destroyed: knives
eaten up by rust, heaters of the tea-urn found in the ash-
hole, &c. And what still further increased my difficulty
was, that my brother's time was entirely taken up with
business, so that I only saw him at meals. Breakfast was
at 7 o'clock or before (much too early for me),—who would
rather have remained up all night than be obliged to rise at
so early an hour. . . .

 "The three winter months passed on very heavily.
I had to struggle against *heimwehe* (home sickness) and low
spirits, and to answer my sister's melancholy letters on the
death of her husband, by which she became a widow with
six children. I knew too little English to derive any
consolation from the society of those who were about
me, so that, dinner-time excepted, I was entirely left to
myself."

D

Introductions to her brother's scholars led to occasional evening parties, where her voice was in demand
as well for single songs as to take part in duets and
glees, and one of these ladies, Mrs. Colebrook, invited
her to go to London on a visit. This visit was
prolonged for several weeks owing to the deep snow,
which rendered the roads impassable. The Duchess of
Ancaster is said to have offered any sum to have a
passage cut near Devizes, but without success, her Grace
was in consequence unable to be present on the 18th
January, when the Queen's birthday was kept. Operas,
plays, auctions, and all the usual amusements of the
town, gave Miss Herschel a glimpse of the gay world;
but the expense of dress and chairmen troubled her
spirits too much to allow of her finding pleasure in
these dissipations; and although Mrs. Colebrook is
allowed to be both "learned and clever," her society
does not appear to have contributed much more to
her happiness than that of some younger ladies whose
companionship was offered, but whose visits she did
not encourage, because, as she bluntly explains, she
"thought them very little better than idiots."

"The time when I could hope to receive a little more
of my brother's instruction and attention was now drawing
near; for after Easter, Bath becomes very empty; only a few of
his scholars whose families were resident in the neighbourhood remaining. But I was greatly disappointed; for, in
consequence of the harassing and fatiguing life he had led

during the winter months, he used to retire to bed with a
bason of milk or glass of water, and Smith's "Harmonics and
Optics," Ferguson's "Astronomy," &c., and so went to sleep
buried under his favourite authors; and his first thoughts
on rising were how to obtain instruments for viewing those
objects himself of which he had been reading. There being in
one of the shops a two and a half foot Gregorian telescope to
be let, it was for some time taken in requisition, and served
not only for viewing the heavens but for making experiments
on its construction. . . . It soon appeared that my brother
was not contented with knowing what former observers had
seen, for he began to contrive a telescope eighteen or
twenty feet long (I believe after Huyghen's description). . .
. . I was much hindered in my musical practice by my help
being continually wanted in the execution of the various contrivances, and I had to amuse myself with making the tube of
pasteboard for the glasses which were to arrive from London, for at that time no optician had settled at Bath. But
when all was finished, no one besides my brother could get
a glimpse of Jupiter or Saturn, for the great length of the
tube would not allow it to be kept in a straight line. This
difficulty, however, was soon removed by substituting tin
tubes. . . . . My brother wrote to inquire the price of a
reflecting mirror for (I believe) a five or six foot telescope.
The answer was, there were none of so large a size, but a
person offered to make one at a price much above what my
brother thought proper to give. . . . . About this time he
bought of a Quaker resident at Bath, who had formerly
made attempts at polishing mirrors, all his rubbish of patterns, tools, hones, polishers, unfinished mirrors, &c., but
all for small Gregorians, and none above two or three
inches diameter.

"But nothing serious could be attempted, for want of
time, till the beginning of June, when some of my brother's

scholars were leaving Bath; and then to my sorrow I saw almost every room turned into a workshop. A cabinet-maker making a tube and stands of all descriptions in a handsomely furnished drawing-room; Alex putting up a huge turning machine (which he had brought in the autumn from Bristol, where he used to spend the summer) in a bed-room, for turning patterns, grinding glasses, and turning eye-pieces, &c. At the same time music durst not lie entirely dormant during the summer, and my brother had frequent rehearsals at home, where Miss Farinelli, an Italian singer, was met by several of the principal performers he had engaged for the winter concerts. . . . . He composed glees, catches, &c., for such voices as he could secure, as it was not easy to find a singer to take the place of Miss Linley. . . . . Sometimes, in the absence of Fisher, he gave a concerto on the oboe, or a sonata on the harpsichord; and the solos on the violoncello of my brother Alexander were divine! . . . . He also took great delight in a choir of singers who performed the cathedral service at the Octagon Chapel, for whom he composed many excellent anthems, chants, and psalm tunes.* As soon as I could pronounce English well enough I was obliged to attend the rehearsals, and on Sundays at morning and evening service, which, though I did not much like at first, I soon found to be both pleasant and useful.

---

* Although a considerable quantity of Sir W. Herschel's musical compositions exist in manuscript, much has unhappily perished. His sister writes:—"I only lament that this anthem was left with the rest of my brother's sacred compositions, which were left in trust with one of the choristers. The morning and evening services each in two different keys, and numerous psalm tunes most beautifully set. The organ book containing the scores; the parts written out and bound in leather, in a box with lock and key which was always kept at the chapel. All is lost. With difficulty many years after, one Te Deum was recovered, and when I was in Bath in 1800 I obtained two or three torn books of odd parts." The chorister's wife openly charged Mr. Linley with having taken possession of these treasures.

"But every leisure moment was eagerly snatched at for resuming some work which was in progress, without taking time for changing dress, and many a lace ruffle was torn or bespattered by molten pitch, &c., besides the danger to which he continually exposed himself by the uncommon precipitancy which accompanied all his actions, of which we had a melancholy sample one Saturday evening, when both brothers returned from a concert between 11 and 12 o'clock, my eldest brother pleasing himself all the way home with being at liberty to spend the next day (except a few hours' attendance at chapel) at the turning bench, but recollecting that the tools wanted sharpening, they ran with the lantern and tools to our landlord's grindstone in a public yard, where they did not wish to be seen on a Sunday morning. . . . But my brother William was soon brought back fainting by Alex with the loss of one of his finger-nails. This happened in the winter of 1775, at a house situated near Walcot turnpike, to which my brother had moved at midsummer, 1774. On a grass plot behind the house preparation was immediately made for erecting a twenty-foot telescope, for which, among seven and ten foot mirrors then in hand, one of twelve foot was preparing; this house offered more room for workshops, and a place on the roof for observing.

"During this summer I lost the only female acquaintances (not friends) I ever had an opportunity of being very intimate with by Bulmer's family returning again to Leeds. For my time was so much taken up with copying music and practising, besides attendance on my brother when polishing, since by way of keeping him alive I was constantly obliged to feed him by putting the victuals by bits into his mouth. This was once the case when, in order to finish a seven foot mirror, he had not taken his hands from it for sixteen hours toge-

ther.*   In general he was never unemployed at meals, but
was always at those times contriving or making drawings
of whatever came in his mind.   Generally I was obliged
to read to him whilst he was at the turning lathe, or polishing mirrors, Don Quixote, Arabian Nights' Entertainment,
the novels of Sterne, Fielding, &c.; serving tea and supper
without interrupting the work with which he was engaged,
. . . . and sometimes lending a hand.   I became in time as
useful a member of the workshop as a boy might be to his
master in the first year of his apprenticeship. . . . . But
as I was to take a part the next year in the oratorios, I had
for a whole twelvemonth two lessons per week from Miss
Fleming, the celebrated dancing mistress, to drill me for a
gentlewoman (God knows how she succeeded).   So we lived
on without interruption.   My brother Alex was absent from
Bath for some months every summer, but when at home
he took much pleasure to execute some turning or clock-
maker's work for his brother."

News from Hanover put a sudden stop for a time to
all these labours.   The mother wrote, in the utmost
distress, to say that Dietrich had disappeared from his
home, it was supposed with the intention of going to
India "with a young idler not older than himself."
His brother immediately left the lathe at which he
was turning an eye-piece in cocoanut, and started for

* " The grinding of specula used to be performed by the hand, no machinery
having been deemed sufficiently exact.   The tool on which they were shaped
having been turned to the required form, and covered with coarse emery and
water, they were ground on it to the necessary figure, and afterwards polished
by means of putty or oxide of tin, or pitch spread as a covering to the same tool
in the place of the emery.   To grind a speculum of six or eight inches in
diameter was a work of no ordinary labour ; and such a one used to be considered of great size."—" *Lord Rosse's Telescopes,*" 1844.

Holland, whence he proceeded to Hanover, failing
to meet his brother as he expected. Meanwhile
the sister received a letter to say that Dietrich was
laid up very ill at an inn in Wapping. Alexander
posted to town, removed him to a lodging, and after a
fortnight's nursing, brought him to Bath, where, on
his brother William's return, he found him being well
cared for by his sister, who kept him to a diet of
" roasted apples and barley-water." Dietrich remained
in England, his brother easily procuring him employment until 1779, when he returned to Hanover, and
shortly afterwards married a Miss Reif. The family
now moved to a larger house, 19, New King Street,*
which had a garden behind it, and open space down
to the river. It is incidentally mentioned, " that
here many interesting discoveries besides the Georgium
Sidus were made."

In preparation for the oratorios to be performed
during Lent, Miss Herschel mentions that she copied
the scores of the " Messiah " and " Judas Maccabeus "
into parts for an orchestra of nearly one hundred performers, and the vocal parts of " Samson," besides
instructing the treble singers, of which she was now
herself the first. On the occasion of her first public
appearance, her brother presented her with ten
guineas for her dress,—

* In this house the Georgium Sidus was discovered, 1781; a volcanic
mountain in the moon, 1783. Here the forty-foot was finished, which re-
vealed two more volcanic mountains in the moon, 1789.

" And that my choice could not have been a bad one I
conclude from having been pronounced by Mr. Palmer (the
then proprietor of the Bath theatre) to be an ornament to
the stage. And as to acquitting myself in giving my songs
and recitatives in the 'Messiah,' 'Judas Maccabeus,' &c.,
I had the satisfaction of being complimented by my friends,
and the Marchioness of Lothian, &c., who were present at the
rehearsals, for pronouncing my words like an Englishwoman."

It is evident that had she chosen to persevere, her
reputation as a singer would have been secure. The
following year she was first singer at the concerts, and
was offered an engagement for the Birmingham Festival,
which she declined, having resolved only to sing in
public where her brother was conductor. At this time
he had repeated proposals from London publishers to
bring out some of his vocal compositions, but with the
exception of " The Echo" catch, none of them ever
appeared in print. Besides the regular Sunday services,
concerts and oratorios had to be prepared for and performed in steady routine, sometimes at Bristol also,
while the poor prima-donna-housekeeper " hobbled on"
with one dishonest servant after another, until Whit
Sunday, 1782, when both brother and sister played
and sung for the last time, in St. Margaret's Chapel.
On this occasion, their last performance in public, the
anthem selected for the day was one of the last compositions, of which mention has been made above.

The name of *William Herschel* was fast becoming
famous, as a writer, a discoverer, and the possessor and

inventor of instruments of unheard-of power.   He was
now about to be released from the necessity of devoting
the time to music which he was eager to give to astronomical science.*   It came about as follows :—

. . . . " He was now frequently interrupted by visitors
who were introduced by some of his resident scholars,
among whom I remember Sir Harry Engelfield, Dr. Blagden, and Dr. Maskelyne.   With the latter he was engaged in
a long conversation, which to me sounded like quarrelling,
and the first words my brother said after he was gone was:
' That is a devil of a fellow.'. . . .

. . . . I suppose their names were not known, or were
forgotten ; for it was not till the year 1782 or 1783 that a
memorandum of the names of visitors was thought of . . . .

. . . . My brother applied himself to perfect his mirrors,
erecting in his garden a stand for his twenty-foot telescope ;
many trials were necessary before the required motions for
such an unwieldy machine could be contrived.   Many
attempts were made by way of experiment against a
mirror before an intended thirty-foot telescope could be completed, for which, between whiles (not interrupting the
observations with seven, ten, and twenty-foot, and writing
papers for both the Royal and Bath Philosophical Societies)
gauges, shapes, weight, &c., of the mirror were calculated,
and trials of the composition of the metal were made.   In
short, I saw nothing else and heard nothing else talked of
but about these things when my brothers were together.
Alex was always very alert, assisting when anything new
was going forward, but he wanted perseverance, and never
liked to confine himself at home for many hours together.
And so it happened that my brother William was obliged to

* He was elected a Fellow of the Royal Society, Dec. 6, 1781.

make trial of my abilities in copying for him catalogues, tables, &c., and sometimes whole papers which were lent him for his perusal. Among them was one by Mr. Michel and a catalogue of Christian Mayer in Latin, which kept me employed when my brother was at the telescope at night. When I found that a hand was sometimes wanted when any particular measures were to be made with the lamp micrometer, &c., or a fire to be kept up, or a dish of coffee necessary during a long night's watching, I undertook with pleasure what others might have thought a hardship . . . . Since the discovery of the Georgium Sidus [March 13, 1781], I believe few men of learning or consequence left Bath before they had seen and conversed with its discoverer, and thought themselves fortunate in finding him at home on their repeated visits. Sir William Watson* was almost an intimate, for hardly a day passed but he had something to communicate from the letters which he received from Sir Joseph Banks and other members of the Royal Society, from which it appeared that my brother was expected in town to receive the gold medal. The end of November was the most precarious season for absenting himself. But Sir William went with him, and it was arranged so that they set out with the diligence at night, and by that means his

* "About the latter end of this month [December, 1779] I happened to be engaged in a series of observations on the lunar mountains, and the moon being in front of my house, late in the evening I brought my seven-feet reflector into the street, and directed it to the object of my observations. Whilst I was looking into the telescope, a gentleman coming by the place where I was stationed, stopped to look at the instrument. When I took my eye off the telescope, he very politely asked if he might be permitted to look in, and this being immediately conceded, he expressed great satisfaction at the view. Next morning the gentleman, who proved to be Dr. Watson, jun. (now Sir William), called at my house to thank me for my civility in showing him the moon, and told me that there was a Literary Society then forming at Bath, and invited me to become a member of it, to which I readily consented."
—*Sir W. Herschel's Journal.* This occurred at a house in River Street, which was soon changed for 19, New King Street.

absence did not last above three or four days, when my
brother returned alone, Sir William remaining with his father.

"Now a very busy winter was commencing; for my
brother had engaged himself to conduct the oratorios con-
jointly with Ronzini, and had made himself answerable for
the payment of the engaged performers, for his credit ever
stood high in the opinion of every one he had to deal with.
(He lost considerably by this arrangement.) But, though
at times much harassed with business, the mirror for
the thirty-foot reflector was never out of his mind, and if a
minute could but be spared in going from one scholar to
another, or giving one the slip, he called at home to see how
the men went on with the furnace, which was built in a room
below, even with the garden.

" The mirror was to be cast in a mould of loam prepared
from horse dung, of which an immense quantity was to be
pounded in a mortar and sifted through a fine sieve. It was
an endless piece of work, and served me for many an hour's
exercise; and Alex frequently took his turn at it, for we were
all eager to do something towards the great undertaking.
Even Sir William Watson would sometimes take the pestle
from me when he found me in the work-room, where he
expected to find his friend, in whose concerns he took so
much interest that he felt much disappointed at not being
allowed to pay for the metal. But I do not think my brother
ever accepted pecuniary assistance from any one of his
friends, and on this occasion he declined the offer by saying
it was paid for already.

" Among the Bath visitors were many philosophical gentlemen who used to frequent the levées at St. James's, when
in town. Colonel Walsh, in particular, informed my brother
that from a conversation he had had with His Majesty,
it appeared that in the spring he was to come with his seven-
foot telescope to the King. Similar reports he received

from many others, but they made no great impression nor
caused any interruption in his occupation or study, and as
soon as the season for the concerts was over, and the mould,
&c., in readiness, a day was set apart for casting, and the
metal was in the furnace, but unfortunately it began to leak
at the moment when ready for pouring, and both my
brothers and the caster with his men were obliged to run
out at opposite doors, for the stone flooring (which ought to
have been taken up) flew about in all directions, as high as
the ceiling. My poor brother fell, exhausted with heat and
exertion, on a heap of brickbats. Before the second casting
was attempted, everything which could ensure success had
been attended to, and a very perfect metal was found in the
mould, which had cracked in the cooling.

" But a total stop and derangement now took place, and
nearly six or seven months elapsed before my brother could
return to the undisturbed enjoyment of his instruments and
observations. For one morning in Passion week, as Sir
William Watson was with my brother, talking about the
pending journey to town, my eldest nephew * arrived to pay
us a visit, and brought the confirmation that his uncle was
expected with his instrument in town. A chaise was at the
door to take us to Bristol for a rehearsal in the forenoon,
of the 'Messiah,' which was to be performed the same evening.
The conductor being still lost in conversation with his friend,
was obliged to trust to my poor abilities for filling the music
box with the necessary parts for between ninety and one hundred performers. My nephew had travelled all night, but we
took him with us, for we had not one night in the week,

---

* George Griesbach, who with the rest of the family settled in England,
where they all did well, their musical talents and connections bringing them
a good deal under the notice of the Court. Mr. G. Griesbach's youngest
daughter, Elizabeth, became the wife of Mr. Waterhouse, of the British
Museum. She died in 1874.

except Friday, but what was set apart for an oratorio either
at Bath or Bristol. Soon after Easter a new organ being
erected in St. James's Church, it was opened with two performances of the ' Messiah;' this again took up some of my
brother's time. . . . .

. . . . The Tuesday after Whit Sunday, May 8th, my
brother left Bath to join Sir William Watson at his father's
in Lincoln's Inn Fields, furnished with everything necessary
for viewing double stars, of which the first catalogue had
just then appeared in the ' Philosophical Transactions.' A
new seven-foot stand and steps were made to go in a
moderate sized box, to be screwed together on the spot
where wanted. Flamsteed's Atlas, in which the stars had
during the winter been numbered, catalogues of double
stars, micrometers, tables, &c., and everything which could
facilitate reviewing objects, had been attended to in the preparation for the journey.

"But when almost double the time had elapsed which my
brother could safely be absent from his scholars, Alex, as
well as myself, were much at a loss how to answer their
inquiries, for, from the letters we received, we could learn
nothing but that he had been introduced to the King and
Queen, and had permission to come to the concerts at
Buckingham House, where the King conversed with him
about astronomy."

It was during his absence at this time that the three
following letters were written and received :—

Dear Lina,—

I have had an audience of His Majesty this morning, and met with a very gracious reception. I presented
him with the drawing of the solar system, and had the
honour of explaining it to him and the Queen. My

telescope is in three weeks' time to go to Richmond, and meanwhile to be put up at Greenwich, where I shall accordingly carry it to-day. So you see, Lina, that you must not think of seeing me in less than a month. I shall write to Miss Lee myself; and other scholars who inquire for me. you may tell that I cannot wait on them till His Majesty shall be pleased to give me leave to return, or rather to dismiss me, for till then I must attend. I will also write to Mr. Palmer to acquaint him with it.

I am in a great hurry, therefore can write no more at present. Tell Alexander that everything looks very likely as if I were to stay here. The King inquired after him, and after my great speculum. He also gave me leave to come to hear the Griesbachs play at the private concert which he has every evening. My having seen the King need not be kept a secret, but about my staying here it will be best not to say anything, but only that I must remain here till His Majesty has observed the planets with my telescope.

Yesterday I dined with Colonel Walsh, who inquired after you. There were Mr. Aubert and Dr. Maskelyne. Dr. Maskelyne in public declared his obligations to me for having introduced to them the high powers, for Mr. Aubert has so much succeeded with them that he says he looks down upon 200, 300, or 400 with contempt, and immediately begins with 800. He has used 2500 very completely, and seen my fine double stars with them. All my papers are printing, with the postscript and all, and are allowed to be very valuable. You see, Lina, I tell you all these things. You know vanity is not my foible, therefore I need not fear your censure. Farewell.

I am, your affectionate brother,

WM. HERSCHEL.

Saturday Morning,
    probably *May* 25.

TO MISS HERSCHEL.

Monday Evening, *June* 3, 1782.

DEAR LINA,—

I pass my time between Greenwich and London agreeably enough, but am rather at a loss for work that I like. Company is not always pleasing, and I would much rather be polishing a speculum. Last Friday I was at the King's concert to hear George play. The King spoke to me as soon as he saw me, and kept me in conversation for half an hour. He asked George to play a solo-concerto on purpose that I might hear him; and George plays extremely well, is very much improved, and the King likes him very much. These two last nights I have been star-gazing at Greenwich with Dr. Maskelyne and Mr. Aubert. We have compared our telescopes together, and mine was found very superior to any of the Royal Observatory. Double stars which they could not see with their instruments I had the pleasure to show them very plainly, and my mechanism is so much approved of that Dr. Maskelyne has already ordered a model to be taken from mine and a stand to be made by it to his reflector. He is, however, now so much out of love with his instrument that he begins to doubt whether it *deserves* a new stand. I have had the influenza, but am now quite well again. It lasted only five or six days, and I never was confined with it. . . . There is hardly one single person here but what has had it.

I am introduced to the best company. To-morrow I dine at Lord Palmerston's, next day with Sir Joseph Banks, &c., &c. Among opticians and astronomers nothing now is talked of but *what they call* my great discoveries. Alas! this shows how far they are behind, when such trifles as I have seen and done are called *great.* Let me but get at it again! I will make such telescopes, and see such things— that is, I will endeavour to do so.

The letter ends abruptly with this sentence, and only one more was written during this momentous interval.

TO MISS HERSCHEL.

July 3, 1782.

Dear Carolina,—

I have been so much employed that you will not wonder at my not writing sooner. The letter you sent me last Monday came very safe to me. As Dr. Watson has been so good as to acquaint you and Alexander with my situation, I was still more easy in my silence to you. Last night the King, the Queen, the Prince of Wales, the Princess Royal, Princess Sophia, Princess Augusta, &c., Duke of Montague, Dr. Heberden, M. de Luc, &c., &c., saw my telescope, and it was a very fine evening. My instrument gave general satisfaction. The King has very good eyes, and enjoys observations with telescopes exceedingly.

This evening, as the King and Queen are gone to Kew, the Princesses were desirous of seeing my telescope, but wanted to know if it was possible to see without going out on the grass, and were much pleased when they heard that my telescope could be carried into any place they liked best to have it. About 8 o'clock it was moved into the Queen's apartments, and we waited some time in hopes of seeing Jupiter or Saturn. Meanwhile I showed the Princesses, and several other ladies who were present, the speculum, the micrometers, the movements of the telescope, and other things that seemed to excite their curiosity. When the evening appeared to be totally unpromising, I proposed an artificial Saturn as an object, since we could not have the real one. I had beforehand prepared this little piece, as I guessed by the appearance of the weather in the afternoon we should have no stars to look at. This being accepted with great pleasure, I had the lamps lighted up which illuminated the picture of a Saturn (cut out in pasteboard) at the bottom of the garden wall. The effect was fine, and so natural that the best astronomer might have been deceived. Their royal highnesses and other ladies seemed to be much pleased with the artifice.

I remained in the Queen's apartment with the ladies till about half after ten, when in conversation with them I found them extremely well instructed in every subject that was introduced, and they seemed to be most amiable characters. To-morrow evening they hope to have better luck, and nothing will give me greater happiness than to be able to show them some of those beautiful objects with which the heavens are so gloriously ornamented.

"Sir William Watson returned to Bath after a fortnight or three weeks' stay. From him we heard that my brother was invited to Greenwich with the telescope, where he was met by a numerous party of astronomical and learned gentlemen, and trials of his instrument were made. In these letters he complained of being obliged to lead an idle life, having nothing to do but to pass between London and Greenwich. Sir William received many letters which he was so kind as to communicate to us. By these, and from those to Alexander or to me, we learned that the King wished to see the telescope at Windsor. At last a letter, dated July 2, arrived from Therese, and from this and several succeeding ones we gathered that the King would not suffer my brother to return to his profession again, and by his writing several times for a supply of money we could only suppose that he himself was in uncertainty about the time of his return.

In the last week of July my brother came home, and immediately prepared for removing to Datchet, where he had taken a house with a garden and grass-plot annexed, quite

E

suitable for the purpose of an observing-place. Sir Wm.
Watson spent nearly the whole time at our house, and he
was not the only friend who truly grieved at my brother's
going from Bath; or feared his having perhaps agreed to no
very advantageous offers; their fears were, in fact, not without reason. . . . . The prospect of entering again on the
toils of teaching, &c., which awaited my brother at home
(the months of leisure being now almost gone by), appeared
to him an intolerable waste of time, and by way of alternative
he chose to be Royal Astronomer, with a salary of £200
a year. Sir William Watson was the only one to whom the
sum was mentioned, and he exclaimed, "Never bought
monarch honour so cheap!" To every other inquirer, my
brother's answer was that the King had provided for him.

Everything was immediately packed for the removal,
and on the 1st of August, when the brothers and sister
walked over to Datchet from Slough (where the coach
passed), they found the waggon, with its precious load
of instruments, as well as household furniture, waiting
to be unpacked. The new home was a large neglected
place, the house in a deplorably ruinous condition, the
garden and grounds overgrown with weeds. For a
fortnight they had no female servant at all; an old
woman, the gardener's wife, showed Miss Herschel
the shops, where the prices of everything, from coals
to butcher's meat, appalled her. But these considerations weighed for nothing in her brother's eyes
against the delight of stables where mirrors could be
ground, a roomy laundry, which was to serve for a
library, with one door opening on a large grass-plot,

where " the small twenty-foot " was to be erected ; he
gaily assured her that they could live on eggs and
bacon, which would cost nothing to speak of now that
they were really in the country !

The beginning of October, Alexander was obliged to return
to Bath. The separation was truly painful to us all, and I
was particularly affected by it, for till now I had not had time
to consider the consequence of giving up the prospect of
making myself independent by becoming (with a little more
uninterrupted application) a useful member of the musical profession. But besides that my brother William would have been
very much at a loss for my assistance, I had not spirit enough
to throw myself on the public after losing his protection.

Poor Alexander ! we had hoped at first to persuade him
to change Bath for London, where he had the offer of the
most profitable engagements, and we should then have had
him near us . . . but he refused, and before we saw him
again the next year he was married.

Much of my brother's time was taken up in going, when
the evenings were clear, to the Queen's Lodge to show the
King, &c., objects through the seven-foot. But when the
days began to shorten, this was found impossible, for
the telescope was often (at no small expense and risk
of damage) obliged to be transported in the dark back to
Datchet, for the purpose of spending the rest of the night
with observations on double stars for a second Catalogue.
My brother was besides obliged to be absent for a week
or ten days for the purpose of bringing home the metal of
the cracked thirty-foot mirror, and the remaining materials
from his work-room. Before the furnace was taken down
at Bath, a second twenty-foot mirror, twelve-inch diameter,
was cast, which happened to be very fortunate, for on the

1st of January, 1783, a very fine one cracked by frost in the tube. I remember to have seen the thermometer 1½ degree below zero for several nights in the same year. . . . . .

. . . . In my brother's absence from home, I was of course left solely to amuse myself with my own thoughts, which were anything but cheerful. I found I was to be trained for an assistant-astronomer, and by way of encouragement a tele-scope adapted for " sweeping," consisting of a tube with two glasses, such as are commonly used in a " finder," was given me. I was " to sweep for comets," and I see by my journal that I began August 22nd, 1782, to write down and describe all remarkable appearances I saw in my " sweeps," which were horizontal. But it was not till the last two months of the same year that I felt the least encouragement to spend the star-light nights on a grass-plot covered with dew or hoar frost, without a human being near enough to be within call. I knew too little of the real heavens to be able to point out every object so as to find it again without losing too much time by consulting the Atlas. But all these troubles were removed when I knew my brother to be at no great distance making observations with his various instruments on double stars, planets, &c., and I could have his assistance immediately when I found a nebula, or cluster of stars, of which I intended to give a catalogue; but at the end of 1783 I had only marked fourteen, when my sweeping was interrupted by being employed to write down my brother's observations with the large twenty-foot. I had, however, the comfort to see that my brother was satisfied with my endeavours to assist him when he wanted another person, either to run to the clocks, write down a memorandum, fetch and carry instru-ments, or measure the ground with poles, &c., &c., of which something of the kind every moment would occur. For the assiduity with which the measurements on the diameter of the Georgium Sidus, and observations of other planets, double stars, &c., &c., were made, was incredible, as may be seen by the various papers that were given to the Royal Society in 1783, which papers were written in the daytime, or when cloudy nights interfered. Besides this, the twelve-inch speculum was perfected before the spring, and many hours were spent at the turning bench, as not a night clear enough for observing ever passed but that some improvements were planned for perfecting the mounting and motions of the various instruments then in use, or some trials were made of new constructed eye-pieces, which were mostly executed by my brother's own hands. Wishing to save his time, he began to have some work of that kind done by a watchmaker who had retired from business and lived on Datchet Common, but the work was so bad, and the charges so unreasonable, that he could not be employed. It was not till some time afterwards in his frequent visits to the meetings of the Royal Society (made in moonlight nights), that he had an opportunity of looking about for mathematical workmen, opticians, and founders. But the work seldom answered expectation, and it was kept to be executed with improvements by Alexander during the few months he spent with us.

The summer months passed in the most active preparation for getting the large twenty-foot ready against the next winter. The carpenters and smiths of Datchet were in daily requisition, and as soon as patterns for tools and mirrors were ready, my brother went to town to have them cast, and during the three or four months Alexander could be absent from Bath, the mirrors and optical parts were nearly completed.

But that the nights after a day of toil were not given to rest, may be seen by the observations on Mars, of which a paper, dated December 1, 1783, was given to the Royal Society. Some trouble also was often thrown away during those nights in the attempt to teach me to re-measure

double stars with the same micrometers with which former
measures had been taken, and the small twenty-foot was
given me for that purpose. . . . . I had also to ascertain
their places by a transit instrument lent for that purpose
by Mr. Dalrymple, but after many fruitless attempts it was
seen that the instrument was perhaps as much in fault as
my observations.

July 8.—I began to use the new Newtonian small sweeper,
(for a description of this instrument see note to Neb. No. 1,
V. class, at the end of the catalogue of first 1000 Neb. and
Cl.), but it could hardly be expected that I should meet
with any comets in the part of the heavens where I swept,
for I generally chose my situation by the side of my brother's instrument, that I might be ready to run to the clock
or write down memorandums. In the beginning of December I became entirely attached to the writing-desk, and
had seldom an opportunity after that time of using my
newly-acquired instrument.

My brother began his series of sweeps when the instrument was yet in a very unfinished state, and my feelings
were not very comfortable when every moment I was
alarmed by a crack or fall, knowing him to be elevated
fifteen feet or more on a temporary cross-beam instead of a
safe gallery. The ladders had not even their braces at the
bottom ; and one night, in a very high wind, he had hardly
touched the ground before the whole apparatus came down.
Some labouring men were called up to help in extricating
the mirror, which was fortunately uninjured, but much work
was cut out for carpenters next day.

That my fears of danger and accidents were not wholly
imaginary, I had an unlucky proof on the night of the
31st December. The evening had been cloudy, but about
ten o'clock a few stars became visible, and in the greatest
hurry all was got ready for observing. My brother, at the

front of the telescope, directed me to make some alteration
in the lateral motion, which was done by machinery, on
which the point of support of the tube and mirror rested.
At each end of the machine or trough was an iron hook,
such as butchers use for hanging their joints upon, and
having to run in the dark on ground covered a foot deep
with melting snow, I fell on one of these hooks, which entered my right leg above the knee.   My brother's call,
"Make haste!" I could only answer by a pitiful cry, "I am
hooked!" He and the workmen were instantly with me,
but they could not lift me without leaving nearly two ounces
of my flesh behind.   The workman's wife was called, but
was afraid to do anything, and I was obliged to be my own
surgeon by applying aquabusade and tying a kerchief about
it for some days, till Dr. Lind, hearing of my accident,
brought me ointment and lint, and told me how to use
them.   At the end of six weeks I began to have some fears
about my poor limb, and asked again for Dr. Lind's opinion:
he said if a soldier had met with such a hurt he would have
been entitled to six weeks' nursing in a hospital. I had,
however, the comfort to know that my brother was no loser
through this accident, for the remainder of the night was
cloudy, and several nights afterwards afforded only a few short
intervals favourable for sweeping, and until the 16th January
there was no necessity for my exposing myself for a whole
night to the severity of the season.

I could give a pretty long list of accidents which were
near proving fatal to my brother as well as myself. To make
observations with such large machinery, where all around is
in darkness, is not unattended with danger, especially when
personal safety is the last thing with which the mind is
occupied; even poor Piazzi* did not go home without get-

* This eminent astronomer made inquiries after Miss Herschel long years
afterwards, as is related in the correspondence. See letter from Sir J.
Herschel, dated Catania, 1824, p. 174.

ting broken shins by falling over the rack-bar, which projects in high altitudes in front of the telescope, when in the hurry the cap had been forgotten to be put over it.

In the long days of the summer months many ten- and seven-foot mirrors were finished; there was nothing but grinding and polishing to be seen. For ten-foot several had been cast with ribbed backs by way of experiment to reduce the weight in large mirrors. In my leisure hours I ground seven-foot and plain mirrors from rough to fining down, and was indulged with polishing and the last finishing of a very beautiful mirror for Sir William Watson.

An account of the discoveries made with the twenty-foot and the improvements of the mechanical parts of that instrument during the winter of 1785, is given with the Catalogue of the first 1000 new nebulæ. By which account it must plainly appear that the expenses of these improvements, and those which were yet to be made in the apparatus of the twenty-foot (which in fact proved to be a model of a larger instrument), could not be supplied out of a salary of £200 a year, especially as my brother's finances had been too much reduced during the six months before he received his *first* quarterly payment of *fifty pounds* (which was Michaelmas, 1782). Travelling from Bath to London, Greenwich, Windsor, backwards and forwards, transporting the telescope, &c., breaking up his establishment at Bath and forming a new one near the Court, all this, even leaving such personal conveniences as he had for many years been used to, out of the question, could not be obtained for a trifle; a good large piece of ground was required for the use of the instruments, and a habitation in which he could receive and offer a bed to an astronomical friend, was necessary after a night's observation.

It seemed to be supposed that enough had been done when my brother was enabled to leave his profession that he might

have time to make and sell telescopes. The King ordered
four ten-foot himself, and many seven-foot besides had been
bespoke, and much time had already been expended on
polishing the mirrors for the same. But all this was only
retarding the work of a thirty or forty-foot instrument, which
it was my brother's chief object to obtain as soon as possible ;
for he was then on the wrong side of forty-five, and felt how
great an injustice he would be doing to himself and to the
cause of Astronomy by giving up his time to making telescopes for other observers.

Sir William Watson, who often in the lifetime of his
father came to make some stay with us at Datchet, saw my
brother's difficulties, and expressed great dissatisfaction.
On his return to Bath he met among the visitors there
several belonging to the Court (among the rest Mde.
Schwellenberg), to whom he gave his opinion concerning
his friend and his situation very freely. In consequence of
this my brother had soon after, through Sir J. Banks, the
promise that £2000 would be granted for enabling him to
make himself an instrument.

Immediately every preparation for beginning the great
work commenced. A very ingenious smith (Campion), who
was seeking employment, was secured by my brother, and
a temporary forge erected in an upstairs room.

It soon became evident that the big, tumble-down
old house, which had been taken possession of with
such eagerness, would not do : the rain came through
the ceilings ; the damp situation brought on ague, and
in June the brother and sister left it for a place called
Clay Hall, Old Windsor. But here again unlooked-for
troubles arose in consequence of the landlady being a

"litigious woman," who refused to be bound to reasonable terms, and at length, on the 3rd of April, 1786,
the house and garden at SLOUGH were taken, and all the
apparatus and machinery immediately removed there.

. . . . And here I must remember that among all this
hurrying business, every moment after daylight was allotted
to observing.   The last night at Clay Hall was spent in
sweeping till daylight, and by the next evening the telescope
stood ready for observation at Slough. . . . . A workman for
the brass and optical parts was engaged, and two smiths
were at work throughout the summer on different parts for
the forty-foot telescope, and a whole troop of labourers were
engaged in grinding the iron tools to a proper shape for the
mirror to be ground on (the polishing and grinding by
machines was not begun till about the end of 1788).   These
heavy articles were cast in town, and caused my brother frequent journeys to London, they were brought by water as
far as Windsor. . . . . At Slough no steady out-of-door
workman for the sweeping handle could be met with, and a
man-servant was engaged as soon as one could be found fit
for the purpose.   Meanwhile Campion assisted, but many
memorandums were put down: " Lost a neb. by the blunder
of the person at the handle."   If it had not been sometimes
for the intervention of a cloudy or moonlight night, I know
not when my brother (or I either), should have got any
sleep ; for with the morning came also his workpeople, of
whom there were no less than between thirty and forty at
work for upwards of three months together, some employed
in felling and rooting out trees, some digging and preparing the ground for the bricklayers who were laying the
foundation for the telescope, and the carpenter in Slough,
with all his men.   The smith, meanwhile, was converting a

washhouse into a forge, and manufacturing complete sets of tools required for the work he was to enter upon. Many expensive tools also were furnished by the ironmongers in Windsor, as well for the forge as for the turner and brass man. In short, the place was at one time a complete workshop for making optical instruments, and it was a pleasure to go into it to see how attentively the men listened to and executed their master's orders ; I had frequent opportunities for doing this when I was obliged to run to him with my papers or slate, when stopped in my work by some doubt or other.

I cannot leave this subject without regretting, even twenty years after, that so much labour and expense should have been thrown away on a swarm of pilfering work-people, both men and women, with which Slough, I believe, was particularly infested. For at last everything that could be carried away was gone, and nothing but rubbish left. Even tables for the use of workrooms vanished : one in particular I remember, the drawer of which was filled was slips of experiments made on the rays of light and heat, was lost out of the room in which the women had been ironing. This could not but produce the greatest disorder and inconvenience in the library and in the room into which the apparatus for observing had been moved, when the observatory was wanted for some other purpose ; they were at last so encumbered by stores and tools of all sorts that no room for a desk or an Atlas remained. It required my utmost exertion to rescue the manuscripts in hand from destruction by falling into unhallowed hands or being devoured by mice.

But I will now return to July, 1786, when my brother was obliged to deliver a ten-foot telescope as a present from the King to the Observatory of Göttingen. Before he left Slough on July 3rd, the stand of the forty-foot telescope stood on two circular walls capped with Portland stone (which, cracking by frost, were afterwards covered with oak) ready to receive the tube. The smith was left to continue to work at the tube, which was sufficient employment cut out for him before he would want farther direction. The mirror was also pretty far advanced, and ready for the polish, for I remember to have seen twelve or fourteen men daily employed in grinding or polishing.

To give a description of the task (or rather tasks) which fell to my share, the readiest way I think will be to transcribe out of a day-book which I began to keep at that time, and called " Book of work done."

July 3.—My brothers William and Alex. left Slough to begin their journey to Germany. Mrs. [Alex.] Herschel was left with me at Slough. By way of not suffering too much by sadness, I began with bustling work. I cleaned all the brass-work for seven and ten-foot telescopes, and put curtains before the shelves to hinder the dust from settling upon them again.

4th.—I cleaned and put the polishing-room in order, and made the gardener clear the work-yard, put everything in safety, and mend the fences.

5th.—I spent the morning in needle-work. In the afternoon went with Mrs. Herschel to Windsor. We chose the hours from two to six for shopping and other business, to be from home at the time most unlikely for any persons to call, but there had been four foreign gentlemen looking at the instruments in the garden, they had not left their names. In the evening Dr. and Mrs. Kelly (Mr. Dollond's daughter) and Mr. Gordon came to see me.

6th.—I put all the philosophical letters in order, and the collection of each year in a separate cover.

*   *   *   *   *

12th.—I put paper in press for a register, and calculated for Flamsteed's Catalogue.

Mem.—When Flamsteed's Catalogue was brought into zones in 1783, it was only taken up at 45° from the Pole, the apparatus not being then ready for sweeping in the zenith.

By July 23rd the whole Catalogue was completed all but writing it in the clear, which at that time was a very necessary provision, as it was not till the year 1789 that Wollaston's Catalogue made its appearance. Many sweeps nearer the Pole than the register of sweeps, which only began at 45°, being made, it became necessary to provide a register for marking those sweeps and the nebulæ discovered in them.

14*th.*—Dr. and Mrs. Maskelyne called here with Dr. Shepherd.

15*th.*—I spent the day with Mrs. Herschel at Mrs. Kelly's. We met Dr. and Mrs. Maskelyne and Dr. Shepherd, Marquis of Huntley, &c., &c., there.

16*th.*—I ruled part of the register of sweeps.

*   *   *   *   *

18*th.*—I spent the whole day in ruling paper for the register ; except that at breakfast I cut out ruffles for shirts. Mr. and Mrs. Kelly and Mrs. Ramsden (Dollond's sister) called this evening. I tried to sweep, but it is cloudy, and the moon rises at half-past ten.

19*th.*—In the evening we swept from eleven till one.

20*th.*—Prince Charles (Queen's brother) Duke of Saxe-Gotha and the Duke of Montague were here this morning. I had a message from the King to show them the instruments.

*   *   *   *   *

I had intended to go on with my Diary till my brother's return, but it would be tedious, so of the rest I shall give only a summary account, and will mention in this place that all what follows would but be the same thing over again ; for the advantage of being quietly at work in the presence of my brother to whom I could apply for information the moment a doubt occurred, never returned again, and often have I been racking my poor brains through a day and a night to very little purpose. I found it necessary to continue my memoranda of " work done " to the last day I had the care of my brother's MS. papers. But I had rather copy a few days more, as they contain the discovery of my first comet, and will serve also to show that I attempted to register all discovered nebulæ, after a precept my brother had left me, as this was necessary for revising the MS. of the catalogue of the first thousand nebulæ, which he expected at his return to find ready for correction from the printers.

22nd.—I calculated all the day for Flamsteed's Catalogue. Lord Mulgrave called this evening. . . . .

23rd.—Received letters from Hanover. Finished calculating for Flamsteed's Catalogue.

The two following short letters were carefully preserved, and, though they contain nothing of importance, they are of interest as being of the very few from the same pen which are not on scientific subjects.

FROM W. HERSCHEL TO CAROLINE HERSCHEL.

HANOVER, *Friday, July* 14, 1786.

DEAR SISTER,—

This morning we arrived safely at Hanover. We are a little tired, but perfectly well in health. We travelled extra post all the way through very bad roads. The post is going out in a very little time, so that I write in a hurry that you might hear from us so much sooner. After a night or two of sleep here (by way of recovery) I shall go on to Göttingen ; but when I have collected my thoughts

better together I will write more. Mamma is perfectly well
and looks well. Jacob looks a little older, but not nearly so
much as I expected. In Sophy [Mrs. Griesbach] there is
hardly any change, but a few white hairs on her head.
John [Dietrich] is just the same as before, his little boy
seems to be a charming creature. Farewell, dear Lina. I
hope we shall see you again in a few weeks. I must finish
for Alexander to write. Adieu once more.

FROM W. HERSCHEL TO CAROLINE HERSCHEL.

[*August*, 1786.]

DEAR LINA,—
     We are still in Hanover, and find it a most agree-
able place. I have been in Göttingen, where Jacob went
along with me, and the King's telescope arrived there in
perfect order. The Society of Göttingen have elected me
a member. We long very much to hear from you, as we
have never had a letter yet. This is the fourth we have
sent you, and we hope you received the former ones. This
day fortnight we have fixed for our setting out from this
place, and be assured that we shall be happy to see *old
England* again, though *old Germany* is no bad place.
Yesterday and the day before I have seen the Bishop of
Osnaburgh and the Prince Edward. If an inquiry should
be made about our return, you may say (I hope with truth)
that we shall be back by about the 24th of August. Adieu,
Lina.

24th.—I registered some sweeps in present time and Pole
distance. Prince Resonico came with Dr. Shepherd to see
the instruments. I swept from ten till one.

*   *   *   *   *

28th.—I wrote part of Flamsteed's Catalogue in the
clear. It was a stormy night, we could not go to bed.

29th.—I paid the smith. He received to-day the plates for the forty-foot tube. Above half of them are bad, but he thinks there will be as many good among them as will be wanted, and I believe he intends to keep the rest till they return. Paid the gardener for four days which he worked with the smith. I registered sweeps to-day. By way of memorandum I will set down in this book in what manner I proceed.

I began some time ago with the last sweep which is booked in the old register (Flamsteed's time and P. D.), viz., 571, and at different times I booked 570, 569, 568, 567, 566, 565. To-day I booked 564 ; 563 is marked not to be registered ; 560 and 561 I was obliged to pass over on account of some difficulty. The rest of the day I wrote in Flamsteed's Catalogue. The storm continued all the day, but now, 8 o'clock, it turns to a gentle rain.

30th.—I wound up the sidereal timepiece, Field's and Alexander's clocks, and made covers for the new and old registers.

31st.—I booked 558, 557, and 554; 556, 555, I was obliged to leave out on account of some difficulty.

Mem.—I find I cannot go on fast enough with the registering of sweeps to be serviceable to the Catalogue of Nebulæ. Therefore I will begin immediately to recalculate them, and hope to finish them before they return. Besides, I think the consequences of registering the sweeps backwards will be bad.

August 1.—I have counted one hundred nebulæ to-day, and this evening I saw an object which I believe will prove to-morrow night to be a comet.

2nd.—To-day I calculated 150 nebulæ. I fear it will not be clear to-night. It has been raining throughout the whole day, but seems now to clear up a little.

1 o'clock.—The object of last night *is a comet.*

3rd.—I did not go to rest till I had wrote to Dr. Blagden

and Mr. Aubert to announce the comet. After a few hours'
sleep, I went in the afternoon to Dr. Lind, who, with Mr.
Cavallo, accompanied me to Slough, with the intention of
seeing the comet, but it was cloudy, and remained so all night.

MISS HERSCHEL TO DR. BLAGDEN.

August 2, 1786.

SIR,—

In consequence of the friendship which I know to
exist between you and my brother, I venture to trouble you,
in his absence, with the following imperfect account of a
comet :—

The employment of writing down the observations when
my brother uses the twenty-foot reflector does not often allow
me time to look at the heavens, but as he is now on a visit
to Germany, I have taken the opportunity to sweep in
the neighbourhood of the sun in search of comets; and
last night, the 1st of August, about 10 o'clock, I found an
object very much resembling in colour and brightness the
27 nebula of the *Connoissance des Temps*, with the difference, however, of being round. I suspected it to be a
comet; but a haziness coming on, it was not possible
to satisfy myself as to its motion till this evening.
I made several drawings of the stars in the field of view
with it, and have enclosed a copy of them, with my observations annexed, that you may compare them together.

August 1, 1786, 9^h 50'. Fig. 1. The object in the centre
is like a star out of focus, while the rest are perfectly distinct,
and I suspect it to be a comet.

10^h 33'. Fig. 2. The suspected comet makes now a
perfect isosceles triangle with the two stars *a* and *b*.

11^h 8'. I think the situation of the comet is now as in
Fig. 3, but it is so hazy that I cannot sufficiently see the
small star *b* to be assured of the motion.

By the naked eye the comet is between the 54 and 53 Ursæ Majoris and the 14, 15, and 16 Comæ Berenices, and makes an obtuse triangle with them, the vertex of which is turned towards the south.

Aug. 2*nd*, 10^h 9′. The comet is now, with respect to the stars *a* and *b*, situated as in Fig. 4, therefore the motion since last night is evident.

10^h 30′. Another considerable star, *c*, may be taken into the field with it by placing *a* in the centre, when the comet and the other star will both appear in the circumference, as in Fig. 5.

These observations were made with a Newtonian sweeper of 27-inch focal length, and a power of about 20. The field of view is 2° 12′. I cannot find the stars *a* or *c* in any catalogue, but suppose they may easily be traced in the heavens, whence the situation of the comet, as it was last night at 10^h 33′, may be pretty nearly ascertained.

You will do me the favour of communicating these observations to my brother's astronomical friends.

I have the honour to be,

Sir,

Your most obedient, humble servant,

CAROLINA HERSCHEL.

August 2*nd*, 1786.
SLOUGH, NEAR WINDSOR.

MISS HERSCHEL TO ALEX. AUBERT, ESQ.

SLOUGH, *August* 2, 1786.

DEAR SIR,—

August 1*st*, in the evening, at 10 o'clock, I saw an object very much resembling (in colour and brightness) the 27 of Mr. Messier's Nebulæ, except this object being round. I suspected it to be a comet; but a haziness came on before I could convince myself of its having moved. I made several figures of the objects in the field, whereof I take the liberty to send the first, that you might compare it with what I saw to-night.

In Fig. 1 I observed the nebulous spot in the centre, a bright red but small star upwards, another very faint white star following, and in the situation as marked in the figure. There is a third star preceding, but exceedingly faint. I suspected several more, which may perhaps appear in a finer evening, but they were not distinct enough to take account of.

In Fig. 2, *August 2nd*, are only the red and its following star: the preceding, in Fig. 1, is partly hid in the rays of the comet, and by one or two glimpses I had, I think it is got before it.

In Fig. 3 I took the comet in the edge by way of taking in the assistance of another star of about the same size and colour as that in the centre.

The only stars I can possibly see with the naked eye which might be of service to point out the place of the comet are 53 and 54 Ursæ Maj., from which it is at about an equal distance with the 14, 15, and 16 Comæ Ber., and makes an obtuse angle with them. I think it must be about 1° above the parallel of the 15 Comæ.

I made these observations with my little Newtonian sweeper, and used a power of about 30: the field is about 1½ degree.

I hope, sir, you will excuse the trouble I give you with my wag [*qy.* vague] description, which is owing to my being a bad (or what is better) no observer at all. For these last three years I have not had an opportunity to look as many *hours* in the telescope.

Lastly, I beg of you, sir, if this comet should not have been seen before, to take it under your protection in regard to A. R. and D. C.

F 2

With my respectful compliments to the ladies, your sisters, I have the honour to be,

Sir,

Your most obedient, humble servant,

CAR. HERSCHEL.

DR. BLAGDEN TO MISS HERSCHEL.

GOWER STREET, BEDFORD SQUARE,
August 5, 1786.

MADAM,—

Mr. Aubert's letter, as well as that with which you favoured me, both arrived safe. The evening was fine on Thursday, but Mr. Aubert was prevented from going to Loam Pit Hill, and I have no opportunity of making astronomical observations here, so that I believe the comet has not yet been seen by anyone in England but yourself. Yesterday the visitation of the Royal Observatory at Greenwich was held, where most of the principal astronomers in and near London attended, which afforded an opportunity of spreading the news of your discovery, and I doubt not but many of them will verify it the next clear night. I also mentioned it in a letter to Paris, and in another I had occasion to write to Munich, in Germany. If the weather should be favourable on Sunday evening, it is not impossible that Sir Joseph Banks and some friends from his house may wait upon you to beg the favour of viewing this phenomenon through your telescope.

Accept my best thanks for your obliging attention in communicating to me the news, and believe me to be, with great esteem,

Your obedient, humble servant,

C. BLAGDEN.

ALEX. AUBERT, ESQ., TO MISS HERSCHEL.

LONDON, *7th August*, 1786.

DEAR MISS HERSCHEL,—

I am sure you have a better opinion of me than to think I have been ungrateful for your very, very kind letter of the 2nd August. You will have judged I wished to give you some account of your comet before I answered it. I wish you joy, most sincerely, on the discovery. I am more pleased than you can well conceive that *you* have made it, and I think I see your *wonderfully clever* and *wonderfully amiable* brother, upon the news of it, shed a tear of joy. You have immortalized your name, and you deserve such a reward from the Being who has ordered all these things to move as we find them, for your assiduity in the business of astronomy, and for your love for so celebrated and so deserving a brother. I received your very kind letter about the comet on the 3rd, but have not been able to observe it till Saturday, the 5th, owing to cloudy weather. I found it immediately by your directions; it is very curious, and in every respect as you describe it. I have compared it to a fixed star, on Saturday night and Sunday night. . . . .

*   *   *   *   *

You see it travels very fast—at the rate of 2° 10′ per day—and moves but little in N. P. D. These observations were made with an equatorial micrometer of Mr. Smeaton's construction, which your brother must recollect to have seen at Loam Pit Hill. I need not tell you that meridian observations with my transit instrument and mural quadrant must have been much more accurate. I give you a little figure of its appearance last night and the preceding night upon the scale of Flamsteed's Atlas Cœlestis [here follows the sketch-figure].

By the above, you will see it will be very near 19 of

Comæ Berenices to-night, and it will be a curious observation if it should prove an occultation of one of the stars of the Comæ. Notice has been given to astronomers at home and abroad of the discovery. I shall continue to observe it, and will give you by-and-by a further account of it. In the meanwhile believe me to be, with much gratitude and regard,

Dear Miss Herschel,

Your most obedient and obliged

humble servant,

ALEX. AUBERT.

P.S.—I was glad to hear to-day, by my friends at our club, that they had seen you last night in good health; pray let me know what news you have of your brother, and when we may expect to see him. I have had twice at Loam Pit Hill his serene highness the Duke of Saxe Gotha, and entertained him, Count Bruhl, and Mr. Oriani (a Milanese astronomer), with your comet last night. My sisters return you many thanks for your kind remembrance, and, with their best compliments, enjoin me to wish you joy.

MISS HERSCHEL TO DIETRICH HERSCHEL.

SLOUGH, *August* 4, 1786.

DEAR BROTHER,—

We received yesterday William's and Alexander's letter, and find that they intend to leave Hanover on the 8th of August, therefore they will not see the contents of this. However, as you have an instrument, I think you are entitled to information of a telescopic comet which I happened to discover on the 1st of August, and which I found, by the observations of the 2nd, to have moved nearly three-quarters of a degree. Last night it was cloudy, but I hope the weather will be more favourable another

night, that we may see a little more of it. I believe you
have a pair of Harris's maps; the place where I saw the
comet is between 53 and 54 Ursæ Maj. and the 14, 15, and
16 Comæ Ber. of Flamsteed's Catalogue. All stars of
Flamst. are in Bode's Cat. to be found, and if you cannot do
without it, I dare say it is to be met with at Hanover. . . .

I found it with a magnifier of about 30, with a field of
about 1½ degree. Now, if you have a piece which is nearly
like this, I would advise you to make use of that in sweeping
all around this place, for it must be, by the time you receive
this letter, at a considerable distance.

When I saw it, it appeared like a very bright, but round,
small nebula.

The first letter I received from Hanover from William
gave us the greatest satisfaction imaginable, for it contained
an account of the good health of all our dear relations. I
hope our dear mother does not grieve too much now they
have left her. I dare say William will pay soon another
visit, and then I will take that opportunity of coming to see
her. Farewell, dear brother; give my best love, &c.

To this period of Miss Herschel's life belongs a folio
manuscript book, written with the utmost neatness,
which she sent with one of her various consignments
of papers to her nephew after her return to Hanover,
and introduced as follows :—

DEAR NEPHEW,—
     This is the fragment of a book which was too bulky
for the portfolio in which I was collecting such papers as I
wished might not fall into any other but your own hands.
They contain chiefly answers of your father to the inquiries
I used to make when at breakfast before we separated each
for our daily tasks.

The information is of a very miscellaneous kind, but matters connected with her special study form the greater part of the questions.   For instance :—

" Given the true time of the transit—take a transit.
   Do the same thing another way.
   To find what star Mercury is nearest.
   Take its place in the Nautical Almanac.
   Another way. . . . .

  *   *   *   *   *

   Time of a star's motion to be turned into space.

  *   *   *   *   *

   To adjust the quadrant when fastened to the telescope.

  *   *   *   *   *

   A logarithm given, to find the angle.
   Oblique spherical triangles."

4th.—I wrote to-day to Hanover, booked my observations, and made a fair copy of three letters.   Made accounts. The night is cloudy.

5th.—I calculated nebulæ all day.   The night was tolerably fine, and I saw the comet.

6th.—I booked my observations of last night.   Received a letter from Dr. Blagden in the morning, and in the evening Sir J. Banks, Lord Palmerston, and Dr. Blagden, came and saw the comet.   The evening was very fine.

7th, 8th.—Booked my observations ; was hindered much by being obliged to find a man to assist the smith.   Dr. Lind and Mr. Cavallo came on the 8th, and Mr. Paradise in the afternoon, but the evening was cloudy.

9th.—I calculated 100 nebulæ. . . . .

10th.—Calculated 100 nebulæ.   The smith borrowed a guinea.   He complains of Turner (the gardener), but we will, if possible, have patience till my brother returns.

11*th*. I completed to-day the catalogue of the first thousand.

12*th*. . . . . calculated 200 nebulæ of the second thousand.

13*th*. Professor Kratzensteine, from Copenhagen, was here to-day. In the evening I saw the comet and swept.

14*th*. . . . . I calculated 140 nebulæ to-day, which brought me up to the last discovered nebulæ, and, therefore, this work is finished.

15*th*. I went up with Mrs. H. to Windsor to pay some bills and to buy several articles against my brother's return.

16*th*. . . . . my brothers returned about three in the afternoon.

It would be impossible for me, if it were required, to give a regular account of all that passed around me in the lapse of the following two years, for they were spent in a perfect chaos of business. The garden and workrooms were swarming with labourers and workmen, smiths and carpenters going to and fro between the forge and the forty-foot machinery, and I ought not to forget that there is not one screw-bolt about the whole apparatus but what was fixed under the immediate eye of my brother. I have seen him lie stretched many an hour in a burning sun, across the top beam whilst the iron work for the various motions was being fixed.

At one time no less than twenty-four men (twelve and twelve relieving each other) kept polishing day and night; my brother, of course, never leaving them all the while, taking his food without allowing himself time to sit down to table.

The moonlight nights were generally taken advantage of for experiments, and for the frequent journeys to town which he was obliged to make to order and provide the tools and materials which were continually wanting, I may say by wholesale.

The discovery of the Georgian satellites caused many breaks in the sweeps which were made at the end of 1786 and beginning of 1787, by leaving off abruptly against the meridian passage of the planet, which occasioned much work, both in shifting of the instrument and booking the observations. Much confusion at first prevailed among the loose papers on which the first observations were noted, and some of them have perhaps been lost; for I remember several configurations of the situation of the satellites having been made by Sir William Watson and Mr. Marsden, and only one could be found. . . . .

    *   *   *   *   **

That the discovery of these satellites must have brought many nocturnal visitors to Slough may easily be imagined, and many times have I listened with pain to the conversation my brother held with his astronomical friends when quite exhausted by answering their numerous questions. For I well knew that on such occasions, instead of renewing his strength by going to rest, that there were too many who could not go on without his direction, among whom I often was included, for I very seldom could get a paper out of his hands time enough for finishing the copy against the appointed day for its being taken to town. But considering that no less than seven papers were delivered to the Royal Society in 1786–1787, it may easily be judged that my brother's study had not been entirely deserted. I had always some kind of work in hand with which I could proceed without troubling him with questions; such as the temporary index which I began in June, 1787. Some years after, the index to Flamsteed's observations, calculating the beginning and ending of sweeps and their breadth, for filling up the vacant places in the registers, and works of that kind, filled up the intervals when nothing more necessary was in hand.

My brother Jacob was with us from April till October, 1787, when he returned to Hanover again. Alexander came only for a short time to give his brother the meeting, Mrs. H. being too ill to be left long alone. (She died in January of 1788.)

Professor Snaidecky often saw some objects through the twenty-foot telescope, among others the Georgian satellites. He had taken lodgings in Slough for the purpose of seeing and hearing my brother whenever he could find him at leisure; he was a very silent man.

My brother's bust was taken by Lockie, according to Sir Wm. Watson's order. Professor Wilson and my brother Jacob* were present.

In August an additional man-servant was engaged, who would be wanted at the handles of the motions of the forty-foot, for which the mirror in the beginning of July was so far finished as to be used for occasional observations on trial.

Such a person was also necessary for showing the telescopes to the curious strangers, as by their numerous visits my brother or myself had for some time past been much incommoded. In consequence of an application made through Sir J. Banks to the King, my brother had in August a second £2,000 granted for completing the forty-foot, and £200 yearly for the expense of repairs, such as ropes, painting, &c., &c., and the keep and clothing of the men who attended at night. A salary of fifty pounds a year was also settled on me as an assistant to my brother, and in October I received twelve pounds ten, being the first quarterly payment of my salary, and the first

---

* This is the last time that the name of Jacob Herschel appears.

money I ever in all my lifetime thought myself to be at liberty to spend to my own liking. A great uneasiness was by this means removed from my mind, for though I had generally (and especially during the last busy six years) been almost the keeper of my brother's purse, with a charge to provide for my personal wants, only annexing in my accounts the memorandum *for Car.* to the sums so laid out—when cast up, they hardly amounted to seven or eight pounds per year since the time we had left Bath. Nothing but bankruptcy had all the while been running through my silly head, when looking at the sums of my weekly accounts and knowing they could be but trifling in comparison with what had been and had yet to be paid in town, for my brother had not been fortunate enough to meet with a reasonable man for a caster who could also furnish the crane, &c., and his bills came in greatly overcharged. But more of this in another place. I will only add that from this time the utmost activity prevailed to forward the completion of the forty-foot. An additional optical workman was engaged, and preparation made for casting the second mirror. Journeys to town were made for moulding, and at the end of January a fine cast mirror arrived safely at Slough. Several seven-foot telescopes were finished and sent off.

The fine nights were not neglected, though observations were often interrupted by visitors. Messrs. Casini, Meehain, Le Genre, and Carochet spent November 26th and 27th with my brother, and saw many objects in the twenty-foot and other instruments.

The catalogue of the second thousand new nebulæ wanted but a few numbers in March to being complete. The observations on the Georgian satellites furnished a paper which was delivered to the Royal Society in May. The 8th of that month being fixed on for my brother's marriage, it may easily be supposed that I must have been fully employed (besides minding the heavens) to prepare everything as well as I could against the time I was to give up the place of a housekeeper, which was the 8th of May, 1788.

END OF RECOLLECTIONS.

CHAPTER III.

LIFE AFTER HER BROTHER'S MARRIAGE.

WITH the second volume of "Recollections" all connected narrative and detailed relation of daily events ceases, and for the ten years from 1788 to 1798 there is not even the journal, which, however, was resumed in the latter year. All has been destroyed. An event so important as her brother's marriage * is only noticed as fixing the date when the "place of a housekeeper" had to be resigned. Miss Herschel lived from henceforth in lodgings, coming every day for her work, and in all respects continuing the same labours as her brother's assistant and secretary as before. But it is not to be supposed that a nature so strong and a heart so affectionate should accept the new state of things without much and bitter suffering. To resign the supreme place by her brother's side which she had filled for sixteen years with such hearty devotion could

---

* Dr. Herschel married Mary, only child of Mr. James Baldwin, a merchant of the City of London, and widow of John Pitt, Esq., by whom she had one son, who died in early youth. She was a lady of singular amiability and gentleness of character. The jointure which she brought enabled Dr. Herschel to pursue his scientific career without any anxiety about money matters.

not be otherwise than painful in any case; but how much more so in this where equal devotion to the same pursuit must have made identity of interest and purpose as complete as it is rare. One who could both feel and express herself so strongly was not likely to fall into her new place without some outward expression of what it cost her—tradition confirms the assumption—and it is easy to understand how this long significant silence is due to the light of later wisdom and calmer judgment which counselled the destruction of all record of what was likely to be painful to survivors.

Her later letters abundantly show that she had learned to love the gentle sister-in-law whom she so pathetically entreats to hold on with her in their common old age, and the journals of her astronomical researches sufficiently prove that her zeal in "*minding the Heavens*" knew no abatement. It was at this period also that she made some of her most important discoveries. Before the end of 1797 she had announced the discovery of eight comets, to five of which the priority of her claim over other observers is unquestioned. A packet, in coarse paper, bearing the superscription, "*This is what I call the Bills and Receipts of my Comets,*" contains some data connected with the discovery of these objects, each folded in a separate paper, and marked "First Comet," "Second Comet," &c., &c. Some of the correspondence on the occasion of her first discovery has already been quoted, and in a note she explains that many of the letters from distinguished men which she received had been given to collectors of autographs. The letter to the Astronomer Royal, announcing the discovery of her second comet, has been preserved, with his answer.

MISS HERSCHEL TO THE REV. DR. MASKELYNE.

DEAR SIR,—

Last night, December 21st, at 7^h 45′, I discovered a comet, a little more than one degree south—preceding β Lyræ. This morning, between five and six, I saw it again, when it appeared to have moved about a quarter of a degree towards δ of the same constellation. I beg the favour of you to take it under your protection.

Mrs. Herschel and my brothers join with me in compliments to Mrs. Maskelyne and yourself, and I have the honour to remain,

Dear sir,

Your most obliged, humble servant,

CAROLINA HERSCHEL.

SLOUGH, *Dec.* 22, 1788.

P.S.—The comet precedes β Lyræ 7′ 5″ in time, and is in the parallel of the small star (β being double). See fifth class, third star, of my catalogue.—WM. H.

THE REV. DR. MASKELYNE TO MISS HERSCHEL.

GREENWICH, *December* 27, 1788.

DEAR MISS CAROLINE,—

I thank you for your favour of the 22nd instant, con-

taining an account of your discovery of a *second* comet
on the 21st, and recommending it to my attention.

I received it only on the 24th, at ten in the morning,
owing to the slowness of our penny post.

I delayed acknowledging it till I could inform you at the
same time I had seen it. The frost, unfortunately for us
astronomers, broke up the very same morning that your
letter arrived, in consequence of which the weather has been
so bad that I could not get a sight of your comet till last
night, the 26th, when, at 6^h 34′, it followed a Lyræ in the
A. R., 3′ 7″ of time, and was 2° 30′ S. of it. This only by
the divisions of the equatorial and meridian circles, but true
to a minute or two of declination and five seconds of time.
I compared it more accurately with a small telescopic star
nearer it, which, when settled hereafter, will determine its
place within 30″ of a degree. Hence its A. R. was about
18^h 33′ 55″, and distance from the North Pole 53° 59′. By
your observation of December 22nd, 5^h 31′ in the morning,
its A. R. was 18^h 35′ 12″, and P. D. 56° 56′. Hence it has
moved retrograde in A. R. about the rate of 17′ of time per
day, and 30′ per day northward in declination, which agrees
nearly with your observation of its approach towards δ Lyræ.
Its motion is fortunately favourable for our keeping sight of
it for some time, which may be very useful, especially if it
should be moving from us, which there is an equal chance
for, as the contrary. It appeared to me very faint, and
rather small, but the air was hazy. By its faintness and
slow motion, it is probably at a considerable distance from
the earth. Time will explain these things. Let us hope
the best, and that it is approaching the earth to please and
instruct us, and not to destroy us, for true astronomers
have no fears of that kind. Witness Sir Harry Englefield's
valuable tables of the apparent places of the Comet of 1661,
expected to return at this time, with a delineation of its

orbit, who, in page 7, speaks of the possibility of seeing a curious and beautiful transit of it over the sun's disk, should the earth and comet be in the line of the nodes at the same time, without *horror* at the thought of our being involved in its immense tail. I would not affirm that there may not exist some astronomers so enthusiastic that they would not dislike to be whisked away from this low terrestrial spot into the higher regions of the heavens by the tail of a comet, and exchange our narrow uniform orbit for one vastly more extended and varied. But I hope you, dear Miss Caroline, for the benefit of terrestrial astronomy, will not think of taking such a flight, at least till your friends are ready to accompany you. Mrs. Maskelyne joins me in best compliments to yourself and Dr. and Mrs. Herschel. If your observation was precise as to the difference of A.R. of the comet and β Lyræ, it may be of use for determining the orbit, especially if the comet should be going off from us. I have not yet examined whether it can be the French comet discovered by M. Messier, on the 26th of last month, which was going from the earth. Its apparent motion must have turned at right angles to its former one, which is possible, but not very probable. I could not see your comet with the night glass, nor would its faintness allow of illuminating the wires.

I remain, dear Miss Caroline,

Your obedient and obliged humble servant,

N. MASKELYNE.

DR. HERSCHEL TO SIR H. ENGLEFIELD.

December 22, 1788.

SIR,—

Your intelligence of the comet I received, but on account of the long time elapsed since the 2nd and 3rd of this month we have not been able to recover the fugitive.

Last night, however, my sister discovered a comet near β Lyræ, which you will find no difficulty to follow as its motion is very slow, and the comet a pretty visible object. We saw it again this morning, and it seems to go towards δ Lyræ, you will see it pass by β Lyræ. It is a much larger object than the nebula near β Lyræ, discovered by Mr. Darquier, of Toulouse (*Connoissance des Temps*, 75).

SIR H. ENGLEFIELD TO DR. HERSCHEL.

PETERSHAM, *December* 25, 1788.

DEAR SIR,—

I am much obliged to you for your account of the comet, and beg you to make my compliments to Miss Herschel on her discovery. She will soon be the great comet finder, and bear away the prize from Messier and Mechain.

The weather yesternight was bad, and to-night I have looked for it, in the moments of fine weather, with a good night-glass, but am not sure that I saw it, though I thought I perceived it about half-way between β and δ Lyræ. The glass I used showed D'Arquier's nebula, though but faintly. Before I could get any other telescope ready, the weather clouded. If you have seen it again, pray be so good as to give me its place when you saw it last, and with what power and light it may be seen. I was going to write to Messier about his comet, but have deferred it, as I would not mention yours without your leave, and could not find it in my heart to write without doing it.

Believe me, dear Sir,
With all the wishes of the season,
Your much obliged and faithful
H. C. ENGLEFIELD.

DR. HERSCHEL TO SIR J. BANKS.

Sir,—

The last time I was in town, you expressed a wish to
see my observations on the comet which my sister, Caroline
Herschel, discovered in the evening of the 21st of last
December, not far from β Lyræ.

As she immediately acquainted the Reverend Dr. Maskelyne and several other gentlemen with her discovery, the
comet was observed by many of them. The Astronomer
Royal in particular having, I find, obtained a very good set
of valuable observations on its path, it will be sufficient if I
communicate only those particulars which relate to its first
appearance, and a few other circumstances that may perhaps
deserve to be noticed.

Dec. 21st, 1788.—About 8 o'clock I viewed the comet
which my sister had a little while before pointed out
to me with her small Newtonian *sweeper*. In my instrument, which was a ten-foot reflector, it had the appearance
of a considerably bright nebula, of an irregular round form,
very gradually brighter in the middle, and about five or six
minutes in diameter. The situation was low, and not very
proper for instruments with high powers.

Dec. 22nd.—About half-after 5 o'clock in the morning I viewed it again, and perceived that it had moved
apparently in a direction towards δ Lyræ, or thereabout. I
had been engaged all night with the twenty-foot instrument,
so that there had been no leisure to prepare my apparatus for
taking the place of the comet; but in the evening of the
same day I took its situation three times. . . . .

In every observation I found the small star which accompanies β Lyræ exactly in the parallel of the comet.

These transits were taken with a ten-foot reflector, and
the difference in right ascension, I should suppose, may be
depended upon to within a second of time. The determi-

nation also of the parallel can hardly err so much as 15 seconds of a degree.

This, and several evenings afterwards, I viewed the comet again with such powers as its diluted light would permit, but could not perceive any sort of nucleus which, had it been a single second in diameter, I think, could not well have escaped me. This circumstance seems to be of some consequence to those who turn their thoughts on the investigation of the nature of comets, especially as I have also formerly made the same remark on one of the comets discovered by Mr. Mechain in 1787, a former one of my sister's in 1786, and one of Mr. Pigott's in 1783, in neither of which any defined, solid nucleus, could be perceived.

I have the honour to remain,

Sir, &c.,

WM. HERSCHEL.

SLOUGH, NEAR WINDSOR,
March 3, 1789.

The third comet was discovered on the 7th January, 1790; the fourth on the 17th April of the same year, during her brother's absence from home. It was announced to Sir Joseph Banks in the following letter:—

April 19*th*, 1790.

SIR,—

I am very unwilling to trouble you with incomplete observations, and for that reason did not acquaint you yesterday with the discovery of a comet. I wrote an account of it to Dr. Maskelyne and Mr. Aubert, in hopes that either of those gentlemen, or my brother, whom I expect every day to return, would have furnished me with the means of pointing it out in a proper manner.

But as perhaps several days might pass before I could

have any answer to my letters, or my brother return, I would not wish to be thought neglectful, and therefore if you think, sir, the following description is sufficient, and that more of my brother's astronomical friends should be made acquainted with it, I should be very happy if you would be so kind as to do it for the sake of astronomy.

The comet is a little more than $3\frac{1}{2}°$ following α Andromedæ, and about $1\frac{1}{2}°$ above the parallel of that star. I saw it first on April 17th, 16^h $24'$ sidereal time, and the first view I could have of it last night was 16^h $5'$. As far as I am able to judge, it has decreased in P. D. nearly $1°$, and increased in A. R. something above $1'$.

These are only estimations from the field of view, and I only mention it to show that its motion is not so very rapid.

I am, &c.,

C. H.

MISS HERSCHEL TO ALEX. AUBERT, ESQ.

SLOUGH, *April* 18, 1790.

DEAR SIR,—

I am almost ashamed to write to you, because I never think of doing so but when I am in distress. I found last night, at 16^h $24'$ sidereal time, a comet, and do not know what to do with it, for my new sweeper is not half finished; and besides, I broke the handle of the perpendicular motion in my brother's absence (who is on a little tour into Yorkshire). He has furnished me to that instrument a Rumboides, but the wires are too thin, and I have no contrivance for illuminating them. All my hopes were that I should not find anything which would make me feel the want of these things in his absence; but, as it happens, here is an object in a place where there is no nebula, or anything which could look like a comet, and I would be much obliged to you, sir, if you would look at the place where the annexed eye-draft will direct you to. My brother has swept that part of the heavens, and has many nebulæ there, but none which I must expect to see with my instrument. I will not write to Sir J. Banks or Dr. Maskelyne, or anybody, till you, sir, have seen it; but if you could, without much trouble, give my best respects and that part of this letter which points out the place of the comet to Mr. Wollaston, you would make me very happy.

I am, dear sir, &c., &c.,

C. H.

SIR JOSEPH BANKS TO MISS HERSCHEL.

Soho Square, *April* 20, 1790.

Madam,—

I return you many thanks for the communication you were so good as to make to me this day of your, discovery of a comet. I shall take care to make our astronomical friends acquainted with the obligations they are under to your diligence.

I am always happy to hear from you, but never more so than when you give me an opportunity of expressing my obligations to you for advancing the science you cultivate with so much success.

Dear Madam,

Your faithful servant,

J. Banks.

ALEX. AUBERT, ESQ. TO MISS HERSCHEL.

London, the 21*st April*, Wednesday, 1790.

Dear Miss Herschel,—

I am much obliged to you for your kind letter. The night before last was cloudy. Last night, or rather this morning, about half-past two, I got up to look for the phenomenon ; it was somewhat hazy. I observed with a common night-glass of Dollond's *a faint something* in a line between α and π Andromedæ, much like a faint star ; it had no coma nor fuzzy appearance. By looking at Flamsteed's Atlas I find no small star there. I was preparing to attack it with a good magnifying power, and to get its place with my Smeaton's equatorial micrometer, but when I was ready a haze came on, and soon after too much daylight, so I can say no more to it as yet. If I saw what you judged a comet, it must have moved but little since you saw it ; it was as large as a star of 7th magnitude, but rather faint. I sent this morning to Dr. Maskelyne : he says he could see nothing *with a good night glass*, but will try again the next fair morning, and after trying he will answer you ; in the meanwhile he begs his best compliments. I will also try again. Pray let me know if you think it was the comet I saw. I have mentioned it to no one but to Mr. Wollaston, who thanks you sincerely, but did not find himself well enough to observe ; he lives in Charter House Square ; direct upon occasion there to the Rev. Francis Wollaston.

You cannot, my dear Miss Herschel, judge of the pleasure I feel when your reputation and fame increase ; everyone must admire your and your brother's knowledge, industry, and behaviour. God grant you many years health and happiness. I will soon pay you a visit, as soon as your brother returns. If I have any instrument you wish to use, it is at your service.

Believe me, &c., &c.,

ALEXANDER AUBERT.

REV. DR. MASKELYNE TO MISS HERSCHEL.

GREENWICH, *April* 22, 1790.

DEAR MISS HERSCHEL,—

*   *   *   *   *

*   *   *   *   *

* * If I misunderstand anything I shall be obliged to you for an explanation. The weather has not permitted me to see anything of the comet yet, but it seems now mending, and I hope to be able to make something of it to-morrow morning. Your second communication, at the same time that it gives me fresh spirits as to the certainty of its being a comet, will certainly assist me in more readily finding it. I feared that your using your new telescope might make that a bright comet to you which might prove but a very faint one, if at all visible, in a common night-glass, which is what we first use to discover a comet with. As soon as I shall have seen it I will send you a line. I sent intelligence of your discovery to M. Mechain, at Paris, last Tuesday, and will send to him your farther communication next Friday. Mr. Maskelyne joins me in best compliments to yourself and Mrs. Herschel, and Dr. Herschel on his return. Dr. Shepherd sent advice of it from me last Tuesday to the Master of Trinity, at Cambridge, who perhaps may convey the agreeable intelligence to your brother.

I remain, dear Miss Herschel,

My worthy sister in astronomy,

Your faithful and obliged humble servant,

N. MASKELYNE.

J. DE LA LANDE TO CAROLINE HERSCHEL.

RUE COLLÈGE ROYAL, le 12 Juillet, 1790.

MA CHÈRE ET S'AVANTE COMMÈRE,—

J'ai reçu avec la plus délicieuse satisfaction la première

lettre dont vous m'avez honoré ; je ne pouvois attribuer votre silence à une timidité que vôtre reputation condamne, mais je l'aurais attribué à mon peu de mérite si vous aviez continué de me refuser une réponse. Vous ecrivez si bien que vous ne pouvez pas avoir à cet égard une excuse légitime.

Vous verrez bientôt M. Ungeschick qui a baptisé vôtre filleule Caroline, dites lui qu'elle se porte beaucoup mieux ainsi que le petite Isaac (je l'ai ainsi nommé en mémoire d'Isaac Newton), pour sa sœur je ne pouvois lui donner un nom plus illustre que le vôtre ; c'est ce que j'ai fait remarquer en annonçant sa naissance dans nôtre *Moniteur* ou *Gazette Nationale* du 31 janvier, je ne pouvois vous donner un compère d'un plus grand mérite que M. Delambre. Il fait actuellement des tables des Satellites de Jupiter qui surpassent de beaucoup celles de M. Wargentin.

Vôtre commère ma nièce calcule des tables pour trouver l'heure en mer par la hauteur du soleil. Mde. du Piery calcule des observations d'éclipses. Pour moi, je suis occupé des étoiles, j'en ai déjà 6,000 ; votre compère Le-Français* y met beaucoup de soin. Nous tâchons tous de seconder vos heureux travaux et ceux de vôtre illustre frère ; nous vous prions tous de recevoir vous même et de lui présenter nos respects.

Remerciez le bien de la complaisance qu'il a eu de m'envoyer la rotation de l'anneau dont j'etois bien curieux. Je suis autant d'attachement que de respect, Savante Miss,

Votre très humble et très
obeissant serviteur,
De. la Lande.

Plusieurs de mes étoiles ont servi à comparer vôtre

---

* M. De la Lande's name was *Jerome Le Français dit de la Lande;* it is to himself, therefore, that he here refers. The letter is addressed "Mlle. Caroline Herschel, Astronome Célèbre, Slough."

comète qui a disparu le 30 juin, mais que M. Messier et
M. Méchain ont suivis sans interruption, jusques dans le
crépuscule.

Je vous prie de demander les bontés de vôtre digne frère
pour M. Ungeschick, qui est un astronome de merite, et qui
a bien du zèle, mais en vous voyant le zèle augmentera.

MISS HERSCHEL TO M. DE LA LANDE.

SLOUGH, *Sept.* 12*th*, 1790.

DEAR SIR,—

Our good friend, General Komavzewski, will persuade
me to believe that I am capable of giving you pleasure by
writing a few lines ; but I am under an apprehension that
he is overrating my abilities.   You, my dear sir, certainly
overrated them when you thought me deserving of expressing your esteem for me in so public a manner as the General
and Mr. Ungeshick have informed me of.

I do not only owe you my sincerest thanks for your good
opinion of me, but my utmost endeavours shall be to make
myself worthy of it if possible.   My good brother has not
been omissive in furnishing me with the means of becoming
so in some respects.   An excellent Newtonian sweeper, of
five-feet focal length, is nearly completed, which, being
mounted at the top of the house, will always be in readiness for observing whenever my attendance on the forty or
twenty-foot telescopes is not required.

I hope the little god-daughter is in good health, and wish
she may grow and give happiness and pleasure to her parents
and uncle.

I beg to present many respectful compliments to the
ingenious ladies you mentioned in your letter.

Mrs. Herschel desires to be remembered to you, sir.   We
do not give up the hopes of seeing you again at Slough,

and are wishing it may not be long before you visit England again.

I remain, dear sir,

With greatest esteem, &c., &c.,

C. HERSCHEL.

Another foreign correspondent was inspired to soar above the ordinary level of scientific communications, and addressed Miss Herschel in a strain of high-flown adulation, of which the following is a translation :—

GÖTTINGEN, *May* 10, [about 1793.]

Permit me, most revered lady, to bring to your remembrance a man who has held you in the highest esteem ever since he had the good fortune to enter the Temple of Urania, at Slough, and to pay his respects to its priestess. I still recall the happy hours passed in England in earlier days of sweet remembrance, and above all, those which I was privileged to spend near you in a society as genial as it was intellectual.

Give me leave, noble and worthy priestess of the new heavens, to lay at your feet my small offering on eclipses of the sun, and at the same time to express my gratitude and deepest reverence. The bearer is a young Mr. Johnston, who has been studying here, and is now returning to England. He is a young man of excellent character, and possessed of unusual capacity and attainments.

May I venture to ask, most honoured Miss, that when you or your brother make any discovery, you will grant me early notice of it, as you once had the kindness to promise to do. You can hardly fail to make them at Slough, where every day is rich in discovery, especially when one of your own subjects—the comets—comes to offer its homage.

How happy should I esteem myself if there were any service I could render you here, most admirable lady astronomer, that I might be permitted to prove how entirely my heart is devoted to you.

Prof. Seyffer.

The fifth comet was discovered December 15th, 1791, and a simple record of the fact is all that the packet devoted to it contains, with the information, " My brother wrote an account of it to Sir J. Banks, Dr. Maskelyne, and to several astronomical correspondents." The discovery of the sixth is treated with equal brevity. " Oct. 7, at 8h. mean time. I discovered a comet, my brother settled its place on the 8th, and I wrote to Sir J. Banks, Dr. Maskelyne, and to Mr. Planta. The letter to Mr. Planta is printed in the Philosophical Transactions."

None of the correspondence in connection with the seventh has been preserved, excepting her own letter announcing its discovery to Sir J. Banks.

MISS HERSCHEL TO SIR JOSEPH BANKS.

Slough, *Nov. 8,* 1795.

Sir,—

Last night, in sweeping over a part of the heavens with my five-foot reflector, I met with a telescopic comet. To point out its situation I transcribe my brother's observations of it from his journal.

*   *   *   *   *

*   *   *   *   *

It will probably pass between the head of the Swan and

the constellation of the Lyre, in its descent towards the sun. The direction of its motion is retrograde.*

     *   *   *   *   *

     *   *   *   *   *

As the appearance of one of these objects is almost become a novelty, I flatter myself that this intelligence will not be uninteresting to astronomers, and therefore hope, sir, you will, with your usual kindness, recommend it to their notice.

I have the honour to be,
With great respect, &c., &c.,
CAROLINE HERSCHEL.

Two years later the eighth and last comet was discovered, on the 6th of August, 1797. It was the occasion of the following letter : —

MISS HERSCHEL TO SIR JOSEPH BANKS.

August, 17, 1797.

SIR,—

    This is not a letter from an astronomer to the President of the Royal Society announcing a comet, but only a few lines from Caroline Herschel to a friend of her brother's, by way of apology for not sending intelligence of that kind immediately where they are due.

I have so little faith in the expedition of messengers of all descriptions that I undertook to be my own, with an intention of stopping in town and write and deliver a letter

---

* This comet, since known as Encke's, in consequence of that great astronomer having determined its periodicity in 1819 and predicted its triennial return, was discovered, independently, four several times before its identity was recognized, Miss Herschel's observation of it in 1795 being the second in order of time. Additional interest has since attached to it, in consequence of its gradually diminishing period and the views hence suggested on the economy of the solar system.

myself, but unfortunately I undertook the task with only the preparation of one hour's sleep, and having in the course of five years never rode above two miles at a time, the twenty to London, and the idea of six or seven more to Greenwich in reserve, totally unfitted me for any action. Dr. Maskelyne was so kind as to take some pains to persuade me to go this morning to pay my respects to Sir Joseph, but I thought a woman who knows so little of the world ought not to aim at such an honour, but go home, where she ought to be, as soon as possible.

The letter which you sent, sir, to my brother, was the only one received at Slough in my absence; it arrived towards noon on the 16th, and was brought by a porter from an inn.

I hope you will excuse the trouble I give by sending this, though I know it is entirely useless, because Dr. Maskelyne had probably my memorandum which I took to Greenwich with him when he called in Soho Square, and therefore I can say nothing but what you, sir, are acquainted with already; but I shall be a little more comfortable when I can say to my brother I have written to Sir J. Banks concerning the comet.

With the utmost respect,

I remain, sir,

Your most obedient servant,

C. Herschel.

We are now reduced to the short diary-like entries in a small book entitled " *Extracts from a Day-Book kept during the years* 1797 *and* 1821," which begins: " 1797, in October I went to lodge and board with one of my brother's workmen (Sprat), whose wife was to attend on me. My telescopes on the roof, to which

I was to have occasional access, as also to the room with the sweeping and observing apparatus, remained in its former order, where I most days spent some hours in preparing work to go on with at my lodging." A chance memorandum shows how the leisure time was employed; thus—"At the ending of 1787, or beginning of 1788, began to make use of some of the proof-sheets of Wollaston's Catalogue along with Flamsteed's;" and again, "December 24th, 1797, received notice for printing the Index, which was not at all adapted for that purpose; but March 8th, 1798, the copy was completed, and taken to the Royal Society, and in the course of the summer the print was corrected." The following letter to the Astronomer Royal bears on this subject:—

MISS HERSCHEL TO REV. DR. MASKELYNE.

SLOUGH, *Sept.* 1798.

DEAR SIR,—

I have for a long while past felt a desire of expressing my thanks to you for having interested yourself so kindly for the little production of my industry by being the promoter of the printing of the Index to Flamsteed's Observations. I thought the pains it had cost me were, and would be, sufficiently rewarded in the use it had already been, and might be in future, to my brother. But your having thought it worthy of the press has flattered my vanity not a little. You see, sir, I do own myself to be vain, because I would not wish to be singular; and was there ever a woman without vanity? or a man either? only with this difference, that among gentlemen the commodity is generally styled ambition.

I wish it were possible to offer something which could be
of use to our Royal Astronomer than merely thanks. Perhaps the enclosed catalogue may be of some little service on
some occasion or other. I was obliged to bring it into that
form by way of scrutinizing the real number of omitted
stars, and find it now very useful when my brother, in
sweeping, &c., observes stars which are not contained in
Wollaston's Catalogue, to know immediately by this order
of R. A. if they are in any of Flamsteed's omitted stars, and
if they are, what number they bear in the catalogue of
omitted stars, which number we find in the first column.
The rest of the columns will want no explanation, except
the last, which would not be complete, or even intelligible,
without the assistance of the catalogue of omitted stars, and
the notes to that catalogue, for they are short memorandums
collected from the descriptions in the catalogue, and from
the notes to some of the stars.

As our Index contains all the corrections and information
which I possibly could collect, those corrections and memorandums of which I had the pleasure, about eighteen months
ago, to write a copy for Dr. Maskelyne, will consequently
be laid aside, else I ought to take notice that there are one
or two errors and several omissions which should have been
corrected in that copy, but with which it will now be needless
to trouble you, sir.

What has laid me under particular obligation to you, my
dear sir, was your timely information, the August before
last, of your having proposed the printing of the Index to
the P. R. S. The papers were then in so incomplete a
state, that it needed each moment which could possibly
be spared from other business to deliver them with some
confidence of their being pretty correct.

Many times do I think with pleasure and comfort on the
friendly invitations Mrs. Maskelyne and yourself have given

H

me to spend a few days at Greenwich. I hope yet to have that pleasure next spring or summer. This last has passed away, and I never thought myself well or in spirits enough to venture from home. If the heavens had befriended me, and afforded us a comet, I might, under its convoy, perhaps have ventured at an emigration. However, I cannot help thinking that I shall meet with some little reward for the denial it has been to me not coming this summer in seeing the improvements Miss Maskelyne has made (more perceptibly) in those accomplishments she seemed to be in so fair a way of attaining when I was there last.

With my best respects and compliments to Mrs. M.,

I remain, with the greatest esteem,

Your most obliged and humble servant,

C. HERSCHEL.

DIARY.

May 29*th* and 30*th*.—Was mostly spent at the Observatory, Professor Vince * being there.

July 30*th*.—My brother went with his family to Bath and Dawlish. I went daily to the Observatory and work-rooms to work, and returned home to my meals, and at night, except in fine weather, I spent some hours on the roof, and was fetched home by Sprat.

*   *   *   *   *

September 11*th*.—Dined at my brother's. Professor Pictet and Dr. Ingenhouse, &c., were there. Cloudy night.

October 7*th*.—Finished the MS. Catalogue of omitted stars for Dr. Maskelyne.

*   *   *   *   *

December 31*st*.—*Mem.* Uncommonly harassed in consequence of the loss of time necessary for going backward and

The Rev. S. Vince, a mathematician and natural philosopher]

forward, and not having immediate access to each book or paper at the moment when wanted.

January 4th.—Spent the evening at my brother's. Sir Wm. Watson * and Mr. Wilson † were there.

February 11th.—My brother went to Bath to make some stay there, having taken a house on Sion Hill.

February 26th.—Mrs. Herschel, Miss Cobet, and the servants left Slough for Bath. Russell, the horse-keeper, and his wife, were, along with me, left in charge of the house, from which I seldom was absent at any other time but to go to dinner at my lodging every day at one o'clock.

March 29th.—The Prince of Orange stepped in to ask some questions about planets, &c.‡

Lord Kirkwall and a gentleman came to see the instruments.

April 1st.—My brother arrived at Slough, and on the 11th he took a paper to the R. S., which he brought with him for me to copy in the clear. The fine nights were spent with sweeping.

*	*	*	*	*

* Sir William Watson, M.D., Knight, F.R.S. from 1770 to 1800, when he resigned. He was one of the first members of the Astronomical Society at its foundation in 1821 under the Presidency of William Herschel. His father, also M.D. and Knight, was the eminent botanist and naturalist. He lived much at Dawlish, where the Herschel family frequently went to stay with him.

† Alexander Wilson, M.D., professor of practical astronomy in the University of Glasgow, and first propounder of that theory as to the cause and nature of the spots on the sun, which was afterwards fully corroborated and worked out by Sir W. Herschel.

‡ The Prince's questions were sometimes of a very remarkable kind. On a previous occasion when he "stept in" with a view to having them answered, and was not so fortunate as to find anyone at home, he left the following memorandum: "The Prince of Orange has been at Slough to call at Mr. Herschel's and to ask him, or if he was not at home to Miss Herschel, if it is true that Mr. Herschel has discovered a new star, whose light was not as that of the common stars, but with swallow tails, as stars in embroidery. He has seen this reported in the newspapers, and wishes to know if there is any foundation to that report.—Slough, the 8th of *August*, 1798.—W. Prince of Orange."

May 14*th*.—Was interrupted in works on account of the Montem.

[*Montem*].—Was visited by Mrs. Owen, the Elds, Linds,* &c., at my lodgings, or wherever they could find me.

June.—Began re-calculating all the sweeps as a constant work for leisure time.

*   *   *   *   *

June 8*th*.—My brother returned. I drank tea with him and Mrs. H., and at seven went home to my lodgings.

*   *   *   *   *

July 15*th*.—Agreed for apartments at Newby's, the tailor, in Slough (Mr. S. and Mrs. B. speaking well of them as sober, industrious people), I am to enter at Michaelmas.

*   *   *   *   *

August 19*th*.—I went to Greenwich to meet some company at Dr. Maskelyne's, and after having spent a week at the R. Observatory, I went with Dr., Mrs., and Miss M. to pay a visit to Sir George Schuckburgh, at Buxted Place, where I left the MS. on the 30th, and arrived at Slough the 31st.

It was so very rarely that Miss Herschel ever slept from home, that this visit was a memorable event in her experience. A small sheet, written by Miss Maskelyne, headed " Journal from the 19th to the 30th of August, 1799," is preserved, with the superscription: " By Miss Maskelyne's memorandum only I found it possible to have any recollection of the occurrences during the eleven days I had intended to spend at Greenwich for the purpose of copying the memoran-

* James Lind, M.D., was a Scotchman, who devoted a considerable amount of his time to astronomical observations.

dums from my brother's second volume of Flamsteed's Observations into Dr. Maskelyne's volume. But the succession of amusements, &c., &c., left me no alternative between contenting myself with one or two hours' sleep per night during the six days I was at Greenwich, or to go home without having fulfilled my purpose."

The journal was enclosed in a letter from Mrs. Maskelyne, which bears pleasant testimony to the agreeable impression which her visitor must have made on the ladies, as well as the astronomer.

BUXTED PLACE, August 30, 1799.

DEAR MISS HERSCHEL,

We thank you for your polite message, are sorry you left Buxted at eight o'clock; hoped you would have taken two dishes of coffee, and not gone till half-past eight, for we were up at seven, to be ready to accompany you to Uckfield.

Margaret has sent the enclosed, and will be glad to hear if it is what you meant; she was writing it when you stopped at the door, but did not venture to open it for fear of disturbing us. Present our compliments to Dr. and Mrs. Herschel. Pray let me know what sort of a journey you have had to your dear sweeper, and accept our love.

I am, dear Miss Herschel,
Your humble servant,
S. MASKELYNE.

The following letter has reference to this visit, and is inserted here, although belonging to a somewhat later date :—

MISS HERSCHEL TO THE REV. DR. MASKELYNE.

January, 1800.

DEAR SIR,

If it was not highly necessary to make you acquainted with the safe arrival of your valuable present at Slough, I might perhaps be a long while before I should think myself sufficiently collected to express the grateful feelings the sight of it occasioned me.   My being pleased at having two such useful and convenient instruments has but very little connection with my present ideas ; and if they had come to me from any other hands but those of the Astronomer Royal, I should use them as occasion required, and think myself much obliged to the giver.   But as it is, I cannot help wishing I were capable of doing *something* to make myself deserving of all these kind attentions.

I feel gratified in particular when I think of the stipulation I was making when you were taking measure of the distance [apart] of my eyes : viz., that if you in future should change in opinion, and not think me worthy of the present, not to bestow it on me.

Mrs. Maskelyne's good-natured looks, and all she said at the time, come now again to my remembrance, and seeing not only the binocular (which I had but a conditional expectation of receiving), but also the night-glass, makes me hope that during the time I had the honour of being in the company of such esteemed friends, I have suffered no loss in their former good opinion of me, which was a circumstance I often feared might have happened ; for I have too little knowledge of the rules of society to trust much to my acquitting myself so as to give hope of having made any favourable impressions.

You see, dear sir, that you have done me more good than you were perhaps aware of : you have not only enabled me to peep at the heavens, but have put me into *good humour* with myself.

With my respectful compliments to Mrs. and Miss Maskelyne,

I remain, with many thanks, Dear sir,

Your much obliged and humble servant,

C. HERSCHEL.

The following is from a friend who took the deepest interest in the career of both brother and sister :—

ED. PIGOTT, ESQ. TO MISS HERSCHEL.

BATH, ST. JAMES'S SQUARE,
April 30, 1799.

MADAM,

It is with much satisfaction that I received through the hands of Dr. Herschel, the valuable publication you are so kind as to send me, and which indeed is the more welcome as I have the volumes of the " Historia Cœlestis," and shall most probably have occasion to use them. Were Flamsteed alive, how cordially would he thank you for thus rendering the labours of his life so much more useful and acceptable to posterity, for he surely little thought that his great work required to be elucidated by an additional folio volume of explanations, errata, and indexes, the advantages of which, by their excellence and accuracy, must every day be more and more acknowledged, and future astronomers, as well as those of the present times will doubtless often be conscious of the merit and obligation you are entitled to.

With many thanks, I remain,

Dear madam,

Your most obedient

EDWD. PIGOTT.

Dr. and Mrs. Herschel, whom I have occasionally the

pleasure of seeing, though by no means so often as I could wish, are well, and desired to be mentioned to you.

August 31*st.*—At six in the evening both my brothers arrived from Bath.   Alexander gave me a call.

September 8*th.*—Professor Vince, his lady, and Alexander came to see me.

October 18*th.*—My brother returned from Bath, but with a violent cough and cold, and was obliged to go to Newbury for change of air and meet Mrs. H., who was there on a visit.

November 19*th.*—The bailiffs took possession of my land-lord's goods, and I found my property was not safe in my new habitation.

  *   *   *   *   *

December 31*st.*—The king had been at the Observatory.

  *   *   *   *   *
  *   *   *   *   *

February 1*st.*—My brother went to Bath.

Mem.—Miss Baldwin [a niece of Mrs. Herschel's] and little John* frequently call on me.

  *   *   *   *   *
  *   *   *   *   *

April 28*th.*—My brother went to town for a fortnight.   I was at the Observatory after he was gone, from ten till two, to select work for me to do at home.

April 29*th.*—From ten till three at the Observatory to make order in the books and MSS.

May 1*st.*—Dined with Dr. Lind.   Fetched my nephew from Mrs. Clark and brought him to his boarding-dame Mrs. Howard, at Eton.   Worked every day some hours at the Observatory.

  *   *   *   *   *

---

* The only child of Dr. Herschel. He afterwards became Sir John Herschel. Miss Herschel was very proud as well as fond of him. He is " my nephew." Dr. Herschel is usually called " my brother," in distinction from all the rest of the family.

May 26*th.*—I went to take leave of my nephew, who entered at Dr. Gretton's School.

*   *   *   *   *

June 23*rd.*—Paid my rent, and gave notice of quitting my apartments at Michaelmas.

June 25*th.*—Began to pack up what I must take to Bath with me, for there I am to go!

June 29*th.*—I dined with Mrs. H. and went with her to the Terrace, where I took leave of my friends at the Lodge. Everything was arranged for my books and furniture to remain at my lodging, to which my brother was to keep the keys. But on receiving information they would be seized along with my landlord's goods by bailiffs, I prepared the same night for their removal, and all was safely lodged in a garret at Mrs. H.'s by July 2 at night.

July 3*rd.*—I left Slough by the nine o'clock Newbury coach, and remained with the Miss Whites [at Newbury] till next morning.

July 4*th.*—At six in the evening I was received at Bath by my brother Alex. and his old housekeeper in a house Mrs. H. had taken for the next winter in Little Stanhope Street. The house had been uninhabited, and the furniture moved into it from the house on Sion Hill by strangers, labourers; the things met me helter-skelter in the passage, some belonging to the drawing-room amongst curry-combs and bridles and other stable utensils. My first care was to make an inventory of the whole, before I let a stranger come into the house, but by the 10th of July I hired a maid of all work to assist me to bring the house into habitable order, and by July 29th I was ready for resuming the work of re-calculating sweeps, or despatching some copying, &c., which was sent me by the coach from Slough, and from the printer in London, my brother being with his family at Tunbridge Wells.

Sept. 10*th.*—I received a box from Slough. My brother was come home, and Alex. went to assist in re-polishing the forty-foot mirror, and left Bath Sept. 15 ; he returned

Oct. 2*nd.*—Some of my time during his absence I spent at his house on Margaret's Hill to clean and repair his furniture, and making his habitation comfortable against his return.

Oct. 29*th.*—I received notice that in about a fortnight I should be wanted at Slough.

FROM DR. HERSCHEL IN LONDON TO CAROLINE HERSCHEL
AT BATH.

LONDON, *Nov.* 7, 1800.

DEAR SISTER,—

Last night my paper on which I have been so long at work was read at the society. I came to London to bring it, and have been so hurried as not to be able to look out any work for you, but shall now be at liberty to do something of that kind. My things here are in considerable disorder, and in a short time Mrs. Herschel and myself wish to come for a little time to Bath, then we will let you know if it's soon, that you may come here on a visit before we go, that I may point out to you the work that is most necessary to be done in our short absence. I thought it best to give you this early notice, because, though we have not fixed upon the time, it will be towards the latter end of this month that we mean to come for perhaps a fortnight or three weeks, according to the weather ; for, if that should be fine we shall return, that I may have a few sweeps before you go back to Bath. Miss Baldwin is at Slough, and stays while we are away, so that you will have company, and the chaise will also be left, so that you can pay visits at Windsor, and show yourself to all your friends and ours.

My last paper consisted of eighty pages, so that you will have a piece of work to gather it together out of the scraps I leave. Some part of it was brought together in the beginning by Miss Baldwin and Mrs. Herschel which will show the order, but the rest remains in bits, which I have gathered together and numbered. . . . .

Remember me to our good brother Alexander, and, with compliments from Mrs. Herschel,

I remain, dear sister,

Your affectionate brother,

Wm. Herschel.

P.S.—The bacon and cheese are very excellent. I have not had time to try Alexander's green lenses; they look beautiful.

Nov. 14*th.*—I left Bath, slept the night at the inn at Newbury, and left there between three and four.

Nov. 15*th.*—I arrived at my brother's house, and as soon as I had dined began to calculate and copy a paper which was to go to the R.S.

Nov. 24*th.*—My brother went with Mrs. H. and Miss Baldwin to Bath, the keys to Obs., &c., were given me to make order and for despatching memorandums which would have employed me for much longer time than it was likely I should be allowed for doing them to my own satisfaction.

Dec. 15*th.*—The family returned, my brother extremely ill, and the next day I had my furniture transported to Windsor, where I had taken a couple of rooms to board and lodge with my eldest nephew, G. Griesbach, and

Dec. 17*th.*—I slept there for the first night.

March 28*th.*—The MSS. and astronomical books in general were removed out of the observatory above stairs and lodged in my brother's library. This alteration proved to be an additional clog to my business (which besides was daily increasing on me) for I lost by this means my workroom and found it very difficult to keep the necessary order among the MSS. *   *   *   *   *

April 20*th.*—Moved from Windsor to a small house at Chalvy, rented from Mr. House, the wood-cutter.

June 9*th.*—My brother went to Bath; by the 25th he was returned.

July 1*st.*—Alexander came from Bath.

July 29*th.*—I went to Slough to take (along with Alex.) care of the house whilst my brother, with his family, were from home.

*　　　*　　　*　　　*　　　*
*　　　*　　　*　　　*　　　*

February 20*th.*—The first time Mrs. Beckedorff's name being mentioned in my memorandums as having dined with her, and the whole party leaving the dining-room on the Princesses Augusta, Amelia, and the Duke of Cambridge coming in to see me.

March 2*nd.*—I went with Mrs. H. and Miss Baldwin to town on a visit to Dr. and Miss Wilson, and went with a party to F. Griesbach's concert at the Opera House. The 4th we returned.

April 7*th.*—I shut my house at Chalvy, and went with my maid to Slough, the latter to supply the place of the servants Mrs. H. took with her to town.

May 6*th.*—My brother went to take a paper to the R. S., and remained there till the 15th.

May 26*th.*—I returned home to Chalvy very ill with a bad leg, having waited too long before I called in assistance.

June 27*th.*—The carriage was sent to take me to Slough. Hitherto work had been daily sent me.

July 13*th.*—My brother, Mrs. H., my nephew John, and Miss Baldwin left Slough to go to Paris.

August 25*th.*—All returned with my nephew danger-

---

* Mrs. Beckedorff was "the sweet little girl of ten or eleven years old" with whom Miss Herschel had exchanged pleasant greetings when they were both taking lessons in dressmaking from Madame Küster, in Hanover, thirty-five years before. (See p. 22.)

ously ill. Going daily for some hours to work at the Observatory, and to receive visitors and letters, had not hastened my recovery, for it required no less than seven months before I could be without the attendance of Dr. Pope.

March 25th.—I moved from Chalvy to Upton.

April 3rd.—Spent the day at Slough. Dr. and Miss Wilson, Miss Whites, and Professor Johnes, from Cambridge, were there.

April 12th.—Had an account of my sister Griesbach's death. She died March 30th.

May 1st.—From the 1st till the 18th I worked with my brother at Slough, when he went to town, and I returned to Upton; but went daily to the library to work till the 26th, when my brother, with his family, came home from town.

June 13th.—Alexander arrived from Bath.

June 25th.—Spent a melancholy day at the Queen's Lodge on account of the French having taken Hanover.

September 18th.—My brother Alex. returned to Bath.

October 18th.—I changed my rooms for the accommodation of Mr. and Mrs. Slaughter, who had taken the house and gardens at Upton, excepting two rooms for my habitation.

November 6th.—I spent the day at Slough. Professor Valis,* with his lady, from Marlow, was there.

November 19th.—I dined at Slough to meet Dr., Mrs., and Miss Maskelyne.

December.—Almost throughout the whole month I worked at Slough from breakfast till nine in the evening.

March 16th.—Finished re-calculating sweeps.

Mem.—Above 8,760 observations have been brought to [the year] 1800.

April 4th.—Dined at Slough to meet Mrs. Bates and a large party. In the evening we heard Mrs. B. sing *Mad Bess*, &c., &c.

---

* Probably Professor Wales, mathematical master at Christ's Hospital, author of a mathematical paper published in the "Phil. Trans.," 1781.

April 18*th.*—I went to Slough. My brother went, with his family, to Bath.

May 10*th.*—My brother returned.

August 5*th.*—My brother Alexander came from Bath.

*   *   *   *   *
*   *   *   *   *

November 22*nd.*—I went to make some stay at Slough during the time my brother spent in town with his family.

December 10*th.*—I returned to Upton.

January 14*th.*—I went, with my brother's family, to a morning concert, to my nephew, H. Griesbach, to hear the Hanoverian Concert-Meister Le Vec play.

*   *   *   *   *

March 5*th.*—Went to make some stay with my brother at Slough, Mrs. H. being in town.

March 27*th.*—All returned, and I went with my work to Upton again.

*   *   *   *   *
*   *   *   *   *

August 14*th.*—I went to stay with Alex. at Slough while my eldest brother went with his family from home. They had intended to have left Slough on the 12th, but were detained in consequence of a report of an expected invasion.

*   *   *   *   *
*   *   *   *   *

In September was much hindered in my work by the packing of the Spanish telescope, which was done at the barn and rick-yard at Upton, my room being all the while filled with the optical apparatus.

September 24*th.*—I went to work with my brother at Slough.

October 1*st.*—When Mrs. H., with her niece, returned from Newbury, I went again to Upton. The Spanish telescope left England in October.*

November 13*th.*—I went to Slough, the family to town;

---

* The cost of this fine instrument, which had been ordered by the King of but, in the absence of the moon, my brother was at home, and much observing, and work was despatched.

December 1st.— All came home, and I went to my solitude again.* During the winter months I suffered much from a violent cough and cold, and found great difficulty in despatching the copying, &c., which daily was sent to me when I was unable to go to my brother.

*   *   *   *   *

May 1st.—I went to Slough to make some stay with my brother.

*   *   *   *   *

July 4th.—My brother went to Gravesend to meet my youngest brother (who came to pay us a visit), and was detained there for a passport.

July 6th.—In the evening they both arrived at Slough.

July 10th.—Alexander joined us from Bath. . . . . The same day my eldest brother went to the visitation of the Observatory at Greenwich, and my brother D. accompanied him. They returned on the 12th.

July 13th.—We went all to the Terrace, and took our tea with Mrs. Bremeyer and Mr. Beckedorff at the Castle.

July 23rd.—Dietrich took leave of his friends at Cumberland Lodge. Alex. and I accompanied him. In Windsor I went shopping to buy presents for my Hanoverian relations.

Spain as long before as January, 1796, was £3150. The Prince of Canino paid £2310 for a ten and a seven-foot telescope from the same indefatigable hands. But although the pecuniary profit was great, it is not surprising that Miss Herschel should bemoan the "making and selling of telescopes" as unworthy of the enormous amount of time and labour which must be withdrawn from the study of astronomy ; and it is evident that the fatigue and exhaustion from polishing mirrors told seriously upon Sir William's health.

* A characteristic little note from her brother belongs to this time : " Lina, —Last night I 'popt' upon a comet. It is visible to the naked eye, between Fomalhout and β Ceti, but above the line that joins the two stars. It made an equilateral triangle (downwards) with 100 and 107 Aquarii. I wrote last night to Sir J. Banks and write now also to Dr. Maskelyne. Adieu.

Dec. 9, 1805."

July 24*th*.—D. left us. My eldest brother and Mrs. H. accompanied him to London.

*   *   *   *   *

*   *   *   *   *

August 1*st*.—I left Upton for Slough. My brother went with Mrs. H. and Miss B. on an excursion. My nephew went to spend the holidays at Newbury, at the Miss Whites. One man and a woman were left with me to take care of the house. I distracted my thoughts by undertaking an amazing deal of work; among the rest, I made catalogues of all books and MSS. my brother's library contained, and arranged them, to the best of my knowledge, according to what the confined room would allow.

September 8*th*.—My brother and family returned, and I went with my works to Upton. Dr. and Miss Wilson were at Slough from September 22nd to September 30th.

Mem.—During September, and the early part of October, many days were spent at Slough in assisting my brother when the 40-foot mirror was re-polishing.

December 28*th*.—I went to see Mrs. Bremeyer, but found she had died ten hours before my arrival at the Castle.

January 15*th*.—My brother went to Bath to see his brother and Sir Wm. Watson.

January 24*th*, 5*th*, and 6*th*.—I spent with my friends at Windsor. My brother returned with a violent cough, added to a nervous headache which it had been hoped would, by change of air, have been removed. My brother brought the place of a comet announced in the papers with him. I had also heard of it at the Castle, and saw it on the 27th at Upton. Next day I had my sweeper carried to Slough, but the nights of the 28th, 29th, and 30th were not clear enough, and I could not find it again till the 31st, when my brother began his observations on it. . . . . .

May 2*nd.*—I left Upton for Slough, to work with my brother. Mrs. H. being in town till

June 18*th.*—Spent the day at Slough, Mr. and Mrs. Watt being there on a visit, and a large party to dinner.

Aug. 13*th.*—I went with Mrs. H. and my nephew to pay a visit to our friends at Cumberland Lodge. My brother, again finding it necessary to recruit his strength by absenting himself for a few days from his work-rooms, had left Slough for Tunbridge Wells just the day before, and at our return we found the Duke of Kent, with the Dukes of Orleans, &c., waiting for us, and my nephew [ætat. 15] and myself showed them Jupiter, the Moon, &c., in the seven-foot.

Aug. 29*th.*—I dined at the Castle. The Queen and Princess Elizabeth honoured me with kind enquiries after the health of my brother, &c. The Princesses Augusta and Mary also came to see me in Miss Beckedorff's room. On coming home the next day, I found my brother had arrived the day before.

Sept. 22*nd.*—In taking the forty-foot mirror out of the tube, the beam to which the tackle is fixed broke in the middle, but fortunately not before it was nearly lowered into its carriage, &c., &c. Both my brothers had a narrow escape of being crushed to death.

* * * * *

* * * * *

Oct. 1*st.*—Received an account and letters announcing a comet.

Oct. 2*nd.*—Saw the comet, visible to the naked eye.

Oct. 4*th.*—My brother came from Brighton. The same night two parties from the Castle came to see the comet, and during the whole month my brother had not an evening to himself. As he was then in the midst of polishing the forty-foot mirror, rest became absolutely necessary after a

I

day spent in that most laborious work; and it has ever been my opinion that on the 14th of October his nerves received a shock of which he never got the better afterwards; for on that day (in particular) he had hardly dismissed his troop of men, when visitors assembled, and from the time it was dark till past midnight he was on the grass-plot surrounded by between fifty and sixty persons, without having had time for putting on proper clothing, or for the least nourishment passing his lips. Among the company I remember were the Duke of Sussex, Prince Galitzin, Lord Darnley, a number of officers, Admiral Boston, and some ladies.

Nov. 3*rd.*—I came home to Upton (Mrs. H. returned from Brighton), but went most days to assist my brother in the polishing-room or library, and from the 10th December to the 22nd I was entirely at Slough going on as above uninterruptedly, Mrs. Herschel being with my nephew, and Miss Baldwin at Newbury with the Miss Whites.

Jan.—Many days at work in the library and workrooms assisting my brother.

Feb. 3*rd.*—When at work in the library the Duke of Cambridge came in. We were obliged to a storm for his visit, as he came in for the shelter.

Feb. 6*th.*—When I came to Slough to assist my brother in polishing the forty-foot mirror, I found my nephew very ill with an inflammatory sore throat and fever.

Feb. 9*th.*—Still very ill; and my brother obliged to go on with the polishing of the great mirror, as every arrangement had been made for that purpose. *Mem.* I believe my brother had reason for choosing the cold season for this laborious work, the exertion of which alone must put any man into a fever if he were ever so strong.

Feb. 10*th.*—From this day my nephew's health kept on mending.

Feb. 19th.—My nephew mending, but my brother not well.

Feb. 26th.—My brother so ill that I was not allowed to see him, and till March 8 his life was despaired of, and by

Mar. 10th.—I was permitted to see him, but only for two or three minutes, for he is not allowed to speak.

Mar. 22nd.—He went for the first time into his library, but could only remain for a few moments.

April 7th.—I went to stay at Slough, my brother going by short stages to Bath, Mrs. H., my nephew, and Miss Baldwin with him.

May 9th.—My brother returned, nearly recovered, but with a violent cold and cough caught on the journey.

May 24th.—I went to Slough to be with my brother till the 31st. In fine nights observing; working in the day-time, and writing a paper on comets, filled up the time, though neither my brother nor myself were well.

June 7th.—Was the Montem, of course much company.

June 13th.—I dined at the Castle to meet Lady and Miss Banks, Mr. De Luc,* &c.

    *   *   *   *   *

July 1st.—Alexander arrived at Slough. *Mem.* We received very distressing accounts from our brother at Hanover.

July 21st till 26th.—My brother was absent, and I was daily at work in the library.

Sept. 5th.—Alexander returned to Bath, leaving his brother far from well. The laborious exertions required for the polishing of the forty-foot mirror, besides the overlooking and directing the workmen out of doors, who were at work on the repairs of the apparatus, during the month of August, had again proved too much for him.

---

* De Luc was a geologist of high reputation; an ardent opponent of Huttonian views.

Oct. 4th.—I went to Slough; my brother, Mrs. H., my nephew, and his cousin, went to Brighton. My brother was absent about a week, during which time I worked as long as I could see in the library, and spent the evenings in booking observations, &c., and such works as could be done within doors.

Nov. 2nd.—My brother went to town, endeavouring to gain some information about my brother Dietrich, who, according to a message from a merchant in town, ought to have by this time been in England.

Nov. 6th.—A letter from Harwich arrived informing us that D. was waiting there for a passport.

Nov. 7th.— D. arrived at Slough, but was obliged to return for his trunk and to show himself at the alien office, and I did not see him till the evening of the 9th.

Dec. 19th.—Dietrich left Slough for lodgings in Pimlico, London. Came with Fr. Griesbach the day before Christmas Day, and returned to town the 26th.

Mem. From the hour of Dietrich's arrival in England till that of his departure, which was not till nearly four years after, I had not a day's respite from accumulated trouble and anxiety, for he came ruined in health, spirit, and fortune, and, according to the old Hanoverian custom, I was the only one from whom all domestic comforts were expected. I hope I have acquitted myself to everybody's satisfaction, for I never neglected my eldest brother's business, and the time I bestowed on Dietrich was taken entirely from my sleep or from what is generally allowed for meals, which were mostly taken running, or sometimes forgotten entirely. But why think of it now!

Jan.—Throughout the whole month I had a cough, my nephew a sore throat and fever. Great flood and stormy weather. The communication between Slough and Upton was very troublesome to me.

Jan. 13*th.*—I spent the day at Slough. Dietrich came for the evening to assist at a concert. I was shocked to see him so much worse, but I was obliged to see him return to town the next morning with Fr. Griesbach. I was prevented by my own illness and the severity of the weather from going to see him in town, and

Feb. 5*th.*—I sprained my ankle in coming home in the evening from Slough, by attempting to walk through the snow in pattens, and my brother was obliged to send me work to Upton, for it was not till a fortnight after, that I could walk again, and I felt the effects of the accident for above three months after.

Mar. 9*th.*—I went to Slough to work with my brother. His family were from home. Much work was done during the time, but the polishing the forty-foot was interrupted on the 24th by the hot weather.

*   *   *   *   *
*   *   *   *   *

Oct. 2*nd.*—Alex left Slough. I was very ill, and had Dr. Pope to attend me.

Oct. 9*th.*—Dismissed Pope and went to Dr. Phips.

Oct. 17*th.*—My nephew went to Cambridge. His mother and Miss Baldwin remained in lodgings at Cambridge.

Nov. 20*th.*—Phips pronounced me out of danger from becoming blind, which he ought to have done much sooner, or rather not to have put me unnecessarily under such dreadful apprehension.

Dec. 6*th.*—Dietrich went to London for the winter.

Joseph Brown sc.

SIR Wᴹ HERSCHEL, Bᵀ

From a Drawing by Lady Gordon,
after a Painting by L.T. Abbott in the National Portrait Gallery.

JOHN MURRAY, ALBEMARLE ST 1876.

CHAPTER IV.

* * * * *
* * * * *
* * * * *

April 29*th*.—My nephew took leave of me, returning to Cambridge.

May 4*th*.—I went to Slough, my brother going to town with Mrs. H. He returned after a short stay, and I remained with him till Mrs. H. came home again. Some of my last days of staying at Slough I spent in papering and painting the rooms I was to occupy in a small house of my brother's attached to the Crown Inn, to which I removed.

July 13*th*.—I went to remain at my brother's house during the time he, with Mrs. H. and Miss Baldwin, went to Scotland.

* * * * *

Sept. 18*th*.—My brother and the family returned, and Dietrich came to Slough, a room being prepared for him in my cottage.

* * * * *

Dec. 1*st*.—Dietrich went to town to enter on his winter engagement.

* * * * *
* * * * *
* * * * *
* * * * *

July 22*nd*.—My brother with his family left Slough on a tour to Edinburgh and Glasgow. I went to his house till they returned, Sept. 18th.

Aug. 6*th*.—Dietrich came to Slough, and I left him to the care of Mrs. Cock, at my habitation.

*   *   *   *   *

*   *   *   *   *

*   *   *   *   *

*   *   *   *   *

May 11*th*.—I went to be with my brother; Mrs. H. went to town for a month.

June 1*st*.—Dietrich came to Slough, disengaged from all business in town to spend the last few weeks he was to be in England with us.

June 12*th*.—Mrs. H. returned from town, and I went home to look to the necessary preparations for Dietrich's precarious (*sic*) journey he was obliged to make through Sweden.

June 27*th*.—My eldest brother went to Oxford, came back the 30th, and Alexander arrived the same day from Bath.

July 8*th*.—Dietrich left us; Alex accompanied him to town.

July 14*th*.—Dietrich left Harwich, and at the end of the month we received a letter dated Gottenburg, July 18, and so far we knew that he was safe, but of receiving any further account we had not the least prospect, for all communication, with Hanover in particular, was cut off.

Sept.—Mrs. Goltermann came to see me, and took a bed at my cottage, I being left alone at my brother's house. The family were at Dawlish with Sir William Watson.

Oct. 5*th*.—My nephew left Slough for Cambridge, with intention of not returning till his studies were ended at the University. The latter end of September Mr. Goltermann received a few lines which came open through France to him, dated September 4, showing that a letter of August 15th had been lost, and that at Helsinförs Dietrich had been robbed of his pocketbook when under examination ; to this accident we were indebted for knowing that he was got home, as he was obliged to write for a duplicate bill of exchange; such letters were, though unsealed, allowed to pass through France.

1813.—The three last months of the preceding year I spent mostly in solitude at home, except when I was wanted to assist my brother at night or in his library.

* * * * *

Jan. 25th.—Congratulatory letters arrived from Cambridge on my nephew's having obtained the Senior Wranglership. He was then contending for another prize, which a few days after he also obtained, so that from the time he entered the University till his leaving he had gained all the first prizes without exception.

* * * * *

* * * * *

March 5th.—Miss S. White, with her maid Sally (one of my nephew's nurses), came to be present at my nephew's twenty-first birthday.

March 7th and *8th.*—I joined the company who dined there on this occasion, and I must not forget that my nephew presented me with a very handsome necklace, which I afterwards sent to my niece Groskopf, when a bride, and I being too old for wearing such ornaments.

* * * * *

March 17th.—My nephew went again to Cambridge to offer himself as candidate for a fellowship, there being three vacant, and at the conclusion of the examination he obtained the first choice of the three.

March 25th.—I went to be with my brother. Mrs.

Herschel and Miss Baldwin followed my nephew to Cambridge to assist him in settling his occasional residence there.

*   *   *   *   *

*   *   *   *   *

May 3*rd.*—I intended to pay a long-promised visit to Mrs. Goltermann, but found my brother too busy with putting the forty-foot mirror in the tube, the carriage having broke down between the polishing-room and the tube. Therefore I postponed my journey till I was sure I should not be wanted at home.

May 10*th.*—I went to London, and met with a friendly reception at Mrs. Goltermann's.

May 11*th.*—I went with Mrs. G. and a Mrs. Kramer to Kensington. I remained with Miss Wilson whilst they paid a charitable visit to the two ladies attendant on the Duchess of Brunswick, who were left in a very distressed situation by the death of their mistress.

The evening we spent at Buckingham House with Mrs. Beckedorff.

May 12*th.*—The forenoon and early part of the afternoon were spent in shopping and visiting, the evening again at Buckingham House, where I just arrived as the Queen and Princesses Elizabeth and Mary, and the Princess Sophia Mathilda of Gloucester, were ready to step into their chairs going to Carlton House, full dressed for a fête, and meeting me and Mrs. Goltermann in the hall, they stopped for near ten minutes, making each in their turn the kindest enquiries how I liked London, &c., &c.

On entering Mrs. Beckedorff's room I found Madame D'Arblay (Miss Burney), and we spent a very pleasant evening.

May 15*th.*—I went to the Exhibition; the evening at Baron Best's, where I met the Beckedorffs. On my return home I found a letter from my brother with Sir William

Watson's direction that I might give them the meeting in town. The next morning I spent a few hours with them, and next day Sir William, with Lady Watson and Miss Jay, called on me in Charles Street. Baron Best also called and brought me the place of a comet from the " Hamburger Zeitungen."

May 18*th*.—I went home and found a great deal of work prepared for me. The evening was spent in sweeping for the comet, but I could not find it, the weather was not clear.

June 14*th*.—I returned to continue my works in the daytime at my own rooms, and the fine evenings assisting my brother when observing, but we were much interrupted by Mrs. H. being seriously ill. She was confined to her room and bed from the 25th of June till the 8th of August before perfectly recovered.

July 24*th*.—Alexander arrived at Slough to spend the summer and work with his brother.

*　　　*　　　*　　　*　　　*
*　　　*　　　*　　　*　　　*
*　　　*　　　*　　　*　　　*

Nov. 13*th*.—I had a call from Miss Joanna Baillie.

Nov. 29*th*.—Mr. Rehberg brought the first letter from our brother Dietrich, dated November 10th, which, though still written with great caution, gave us, after a lapse of sixteen months, the assurance that he and his family were living.

Dec. 4*th*.—I met Madame D'Arblay and Mr. Rehberg, &c., at the Castle.

Jan. 1*st*.—My nephew, John Herschel, brought me, for a New Year's present, a new publication by him.

Mem.—The winter was uncommonly severe. My brother suffering from indisposition, and I, for my part, felt I should never be anything else but an invalid for life, but which I very carefully kept to myself, as I wished to be useful to my brother as long as possibly I could. . . . .

Feb. 7*th.*—I was obliged to move to a small cottage in Slough, at a considerable distance from my brother. I began to move, and slept there for the first night, the 22nd.

April 1*st.*—My brother went to Bath to see his brother and Sir William Watson. His cough still very bad, and the 12th, when he came home, we learned that he had been taken very ill on the road and suffered much when at Bath. It was not till many weeks after, when the warm weather came on, that he felt relieved. A few days after his return from Bath, we received notice by a message from the Queen of the Duchess of Oldenburg's intention of coming to see my brother's instruments. Everything was put in readiness for either a morning or evening visit, but the weather being very bad, the visit was put off till the arrival of the Emperor.

May 4*th.*—I went to be with my brother. Mr. H. and Miss B. went to meet my nephew in town, who was keeping a term in the Temple, where he had commenced to be a student for the law in February.

*   *   *   *   *

June 10*th.*—My brother, being about this time engaged with re-polishing the forty-foot mirror, it required some time to restore order in his rooms before any strangers could be shown into them, and I again was assisting him to prepare for the reception of the Emperor Alexander and the Duchess of Oldenburg, &c., as they were at Windsor for Ascot Races. But we might have saved ourselves the trouble, for they were sufficiently harassed with public sights and festivities.

*   *   *   *   *

Sept. 13*th.*—During the time I was with my brother I saw among the visitors, &c., General P., who informed us of General Komarzewsky's death, and on my expressing a hope it might not be true, le General said he had buried him himself at Paris, and had erected to him a little monument as long as seven years ago.

Sept. 30*th.*—I came to my home again, but under the greatest concern at being obliged to leave my brother without my little help. But I have since been with him every morning till he told me he should leave off. His strength is now, and has for the last two or three years not been equal to the labour required for polishing forty-foot mirrors. And it was only by little excursions and absence from his workrooms, he for some time recovered from the effects of over-exertion.

Nov. 15*th.*—I went to work with my brother, which chiefly consisted of calculations and constructing new tables for the Georgian satellites, &c., &c.

Nov. 29*th.*—Mrs. H. returned, and I continued calculating and copying at home.

* * * * *

Aug. 11*th.*—Alexander left Slough, my eldest brother with him, going on to Dawlish to recruit his strength again. His declining health had a sad effect on Alexander's spirits, and I was in continual fear of the consequences; for nothing but the thoughts of the yearly meeting had till now kept up his spirits. From what is yet to follow, it will be seen that our next meeting was not only the last, but a very distressing one.

Sept. 11*th.*—I went to be with my brother, and remained with him till the 12th of October. The first fortnight of my being with him he was not able to do anything which required strength.

* * * * *
* * * * *

Jan. 2*nd.*—I was obliged to attend at Slough by eight

o'clock, to be present when the Archdukes John and Louis of Austria visited my brother and his instruments.

Jan. 9th.—My nephew received a diploma of being Member of the University of Gottingen. The packet brought very satisfactory letters from our brother at Hanover.

Feb. 4th.—My brother sent the carriage to fetch me home [from the Castle], and I was desired to write to our brother Alexander at Bath, from whom a most melancholy letter had that morning arrived, acquainting us with his being confined to his bed, having received an injury to his knee.

April 5th.—My brother received the Royal Hanoverian Guelphic Order.

May 12th.—My brother went to town to prepare for going to a levée at the Regent's next Tuesday. He brought me the keys to the library for going there to work.

* * * * *

June 17th.—I went to my brother's house, and was left in the deepest concern for his health. He went with his family to Cambridge. [Alexander was to make a journey to Hanover.]

Sept. 2nd.—I saw Alexander led by Captain Stevens on board . . . . of whom I had the assurance that he would see Alexander safe to Dietrich's friend, Mr. Münter, in Bremen. A few hours after I left the place [Wapping], taking with me receipts from everybody with whom I had had occasion to keep accounts. I came very ill to Mrs. Goltermann's, where I remained a week under her care.

Sept. 9th.—I went home.

* * * * *

Sept. 23rd.—We were at a fête the Queen gave at Frogmore. I was obliged to return with my brother soon after he had been noticed by and conversed with the Queen and Regent, being too feeble to be long in company.

Sept. 26th.—We had letters from Hanover to acquaint us

with Alexander's arrival in improved health, after a pleasant journey both by sea and land.

October, Nov., Dec.—Nothing particular happened, my nephew remaining at home working with his father, and I took the opportunity of working on my MS. Catalogue at those times when I was left without employment.

*   *   *   *   *

March 27*th.*—I spend the day at my brother's, Sir Robert and Lady Liston being there on a visit before their return to Constantinople.

May 10*th.*—I met Sir William and Lady Watson at dinner at my brother's, but was grieved to see the sad change in Sir William's health and spirits, and felt my only friend and adviser was lost to me.

*   *   *   *   *

June 9*th.*—All the family came home. I returned to my house with astronomical work to finish.

June 14*th.*—Spent the day at Lady Herschel's to meet Mrs. and Miss Maskelyne.

July 10*th* and 11*th.*—Spent at my brother's, the mornings at work in the library the evenings with the company. . . .

July 14*th.*—I spent with Mrs. Beckedorff and brought tickets of invitation to a fête at Frogmore, for our family, with me ; where we all went on the 17th of July ; but almost as soon as the Royal party sat down to dinner I was obliged to go home with my brother, after having twice been honoured by the notice and conversation of the Queen and Regent, &c., &c. He found himself too feeble to remain in company. It was said that there were above two thousand persons invited.

*   *   *   *   *
*   *   *   *   *

Nov. 7*th.*—Prepared for going into mourning for the

Princess Charlotte. Mrs. De Luc died a few days after or before the Princess.

Feb. 11*th.*—I went to my brother, and remained with him till the 23rd. We spent our time, though not in idleness, in sorrow and sadness. He is not only unwell but low in spirits.

April 13*th.*—Princess Elizabeth of Hesse Homburg and the Prince of Hesse Homburg came to see my brother and his instruments. They were attended by Count O———, Baron K———, and Baron G———. The latter being well informed in the science of astronomy.

Mem. I lost my attendants, the C.'s, at the latter end of April, and a waste of my time was the consequence, for I never after met with anyone who was deserving of my trust.

June 8*th.*—The Prince and Princess Schaumburg von der Lippe, attended by Fraulein U., came to see my brother. Their behaviour to him was truly kind and affectionate on leaving him, with a hope to see him in the same place—in the garden at the foot of the forty-foot telescope—five or six years hence, when they should come to England again.

*   *   *   *   *

June 25*th.*—From this day to July 8th I was with my brother. The family at Newbury; he being so far well that without interruption, I was supplied with copying as he wrote.

July 16*th.*—I went to my brother's, to be present in the evening when the Archduke Michael of Russia, with a numerous attendance, came to see Jupiter, &c.

July 21*st.*—

Mem. Began to copy the numbering of stars from my brother's 2nd volume of Flamsteed's Observations into one of my own, having succeeded to procure all the three volumes c omplete at the price of four guineas.

Aug. 8th.—I spent the afternoon with my brother, who found himself very unwell, but with the assistance of my nephew, he had the pleasure of showing the Princess Sophia of Gloucester (who came in the evening accompanied by the Archbishop of Canterbury and several lords and ladies) many objects in the ten-foot telescope.

Aug. 18th.—I went to my brother's, his family left home for Brighton, where he intended to follow as soon as the repairs of the forty-foot should be finished; but he was all the time too ill for being anywhere but at home. The first evening we were alone, the Princess Sophia came to see the moon. She was accompanied by Lady Mary Paulet, another lady, and some gentlemen. After their departure, my brother seemed much pleased with the intelligent enquiries made by the Princess; but with much concern I saw that he had exerted himself too much above his strength.

Aug. 25th.—I was obliged to leave my brother for a few hours to call on the Princess Sophia Matilda, who desired to see me.

*   *   *   *   *

Sept. 8th.—I spent some hours with the Princess at the Castle.

Oct. 14th.—The Ertz Herzog Maximilian of Austria came to see my brother, charged with messages from his mother to both my brother and myself, we having had the honour of seeing her Imperial Highness at Slough, in 1786, when on a visit to the King, with her husband the Archduke of Milan.

Nov. 12th.—I spent some hours in the forenoon with the Princess at the Castle. I left her with a promise of coming soon again, but it was to be my last visit for a long time to come, for . . . .

Nov. 17th.—The Queen died. The 3rd of December

the Princess returned my books with a kind note, and on the 4th she left Windsor.

Dec. 5th, 6th, 7th.—I spent in Windsor to see Mrs. and Miss Beckedorff at short intervals. Miss Wilson, Miss S. White, Miss Baldwin, Mr. Beckwith (Miss B.'s bridegroom) were visitors for several days at Slough, to see the funeral of the Queen.

    *   *   *   *   *

Dec. 16th.—My brother went to town, to sit for his portrait by Mr. Artaud.

Feb. 3rd.—My brother went to town. The 4th I received a note from Mrs. Beckedorff, desiring me to spend the next and last day with her, but I went immediately and took (as I then thought) my last leave of both mother and daughter, for I could not leave my brother on his return on the 5th to be received only by the servants, as he went from home very unwell with a cold.

Feb. 7th.—My nephew arrived in town, and on the 12th all came home and I returned to my habitation.

Feb. 28th.—I heard of the death of Mrs. Beckedorff's daughter, at Hanover. My brother consented to my going next morning to London, and before two o'clock, after I had procured a lodging in Pimlico, I was with the poor mourners at Buckingham House, and remained till March 4th, when I left them, hoping they would be able to leave England on the 9th.

March 11th.—Was Miss Baldwin's wedding-day, which I spent at Slough, with the family.

April 2nd.—My brother left Slough, accompanied by Lady H. for Bath, he being very unwell, and the constant complaint of giddiness in the head so much increased, that they were obliged to be four nights on the road both going and coming.

The last moments before he stepped into the carriage

            K

were spent in walking with me through his library and workrooms, pointing with anxious looks to every shelf and drawer, desiring me to examine all and to make memorandums of them as well as I could. He was hardly able to support himself, and his spirits were so low, that I found difficulty in commanding my voice so far as to give him the assurance he should find on his return that my time had not been misspent.

When I was left alone I found that I had no easy task to perform, for there were packets of writings to be examined which had not been looked at for the last forty years. But I did not pass a single day without working in the library as long as I could read a letter without candlelight, and taking with me papers to copy, &c., &c., which employed me for best part of the night, and thus I was enabled to give my brother a clear account of what had been done at his return.

May 1st.—But he returned home much worse than he went, and for several days hardly noticed my handiworks.

*   *   *   *   *

June 21st.—I went with my brother to town. He was to sit to Mr. Artaud. We remained till Friday, whilst Lady Herschel entertained the Wilson family at home, who were attending the funeral of Miss Wilson at Upton.

July 8th.—We thought my brother was dying. On the 9th he was persuaded to be blooded in the arm which something relieved him.

Aug. 10th.—My brother and Lady H. took me with them to town.

Aug. 11th.—We went to the Bank and did what was thought necessary.

Aug. 12th.—I went with Lady H. to see my brother's portrait, and ordered a copy for myself.

Aug. 25th.—

Mem.—The 13th we came home, and one day passes like

the other. I have much to do and can do but little beyond going daily to my brother, and often we are both unable to look about business. The present hot weather bears hard on enfeebled constitutions. Thermometer most days above 80 degrees.

Oct. 15*th.*—I went to my brother, his family being in town.

Oct. 29*th.*—I returned to my home.

A small slip of yellow paper, containing the following lines, traced by a tremulously feeble hand, belongs to this year :—

" LINA,—There is a great comet. I want you to assist me. Come to dine and spend the day here. If you can come soon after one o'clock we shall have time to prepare maps and telescopes. I saw its situation last night—it has a long tail."

    July 4*th*, 1819.

Then follows :—

" I keep this as a relic ! Every line *now* traced by the hand of my dear brother becomes a treasure to me.

                " C. HERSCHEL."

The next year opens, as so many previous ones have done. The bare facts of the steadily narrowing life being set down with the same brevity and unswerving attention to *the one* object. The family was in much anxiety on account of the failing health of Mrs. Beckwith, the niece of Lady Herschel, of whom, as Miss Baldwin, frequent mention has been made. The spring and summer were passed in taking the sufferer to different places in the country, but she was sinking in a rapid decline, and died in the autumn.

Nov. 10*th.*—The remains of Mrs. Beckwith were brought to Upton to be buried, and to me was left the melancholy task of keeping up my poor brother's spirits on such a melancholy occasion, when at the same time my own were at their lowest ebb, and being besides much molested about this time by the rejoicing of an unruly mob at the acquittal (as they called it) of the Princess of Wales.

From the 26th to 29th I was with my brother.

March.—We lost our brother Alexander, who died at Hanover.*

*   *   *   *   *

May 22*nd.*—Again with my brother. My chief care was to see that my brother was not fatigued by too many visitors, and reading to him to prevent his sleeping too much.

*   *   *   *   *

*   *   *   *   *

The volume ends in October :—1821.

"Here closed my Day-book, for one day passed like

* The following notice is from a Bristol paper :

"Died, March 15th, 1821, at Hanover, Alexander Herschel, Esq., well-known to the public of Bath and Bristol as a performer and elegant musician ; and, who for forty-seven years was the admiration of the frequenters of concerts and theatres of both those cities, as principal violoncello.

"To the extraordinary merits of Mr. Herschel was united considerable acquirement in the superior branches of mechanics and philosophy, and his affinity to his brother, Sir William Herschel, the illustrous astronomer, was not less in science than blood. To a large circle of professional friends the uniform gentlemanly manners of Mr. Herschel have rendered him at once an object of their warmest regard and respect." Alexander Herschel returned to Hanover in September, 1816, and was enabled to live in comfortable independence until his death at the age of seventy-six, through the never failing generosity of his elder brother.

another, except that I, from my daily calls, returned to my solitary and cheerless home with increased anxiety for each following day."

On the 25th of August, 1822, Sir William Herschel died in his house at Slough.

A small book, containing a very few pages, entitled " Memorandum from 1823 to," &c., gives the sad history of the last days of that long life of indefatigable toil over which the devoted sister had watched so long with untiring love. It would be easy, and perhaps in some respects preferable, to tell the story without the details, but it would be at the cost of much that is characteristic and illustrative of the nature which has thus far been unfolded from within, and it is the last chapter of her life which she thought worth recalling to memory and committing to paper. The terrible blow of the death of her brother seems to have deprived her of all power or desire to do or to will anything beyond the one stern, dogged resolve to leave England for ever as soon as the beloved remains were buried from her sight. Six months after her return to Hanover she thus prefaced this last and most pathetic of her *Recollections :—*

HANOVER, *April* 15*th,* 1823.

" Eighteen months have elapsed since I could acquire fortitude enough for noting down in my Day-book any of those heartrending occurrences I witnessed during the last nine months of the fifty years I have lived in England, and I cannot hope that ever a time will come when I shall be able to dwell on any one of those interesting but melancholy hours I spent with the dearest and best of brothers. But if I was to leave off making memorandums of such events as either affect or are interesting to me, I should feel like what I am, viz., a person that has nothing more to do in this world.

"But to regain the thread of my narration, it is necessary to take notice of the vacancy between the present date and the ending of the year 1821, and the only way in which I can possibly fill up this vacancy must be to take a few dates with memorandums marked in my almanac and account books for the year 1822, without making any comments on what my feelings and situation must have been throughout that whole interval.

"By some letters I wrote during the first four months of 1822 to my brother here at Hanover, I see that I was employed in copying from the Philosophical Transactions the first twelve papers of my brother's publications. The time required for this purpose I could only obtain by making use of most of the hours which are generally allotted to rest, as during the day my time was spent in endeavours to support my dear brother in his painful decline. And besides, the hope that we might continue yet a little longer together began to forsake me, for my own health and spirits were in that state that I was in daily expectation of going before.* Therefore each moment of separation from my dear brother I spent in endeavours to arrange my affairs so that my nephew, J. Herschel, as the executor of my will, might have as little trouble as possible.

---

* Although Miss Herschel was endowed by nature with a fine healthy constitution, she suffered much in various ways during the last twenty-five years of her life ; and there is little doubt that her health was injured, to a considerable extent, by the excessive fatigue and serious accidents to which she was exposed in her earlier days, when she often denied herself rest that was imperatively needed, in order to be at hand when her brother required her services.

[A letter of eighteen pages would have been found along with a will, if I had (as I then daily expected) died before my brother. After the sad events of the succeeding two years, I thought it necessary to destroy both the will and the letter.] My thoughts were continually divided between my brother's library, from which I was now on the point of being severed for ever, and my own unfinished work at home endeavouring to bring by degrees all into its proper place."

Diary—(continued).

May 13*th.*—Lady Herschel and my nephew went to town : I was left with my brother alone, but was counting every hour till I should see them again, for I was momentarily afraid of his dying in their absence.

May 20*th.*   *   *   *   *

The summer proved very hot; my brother's feeble nerves were very much affected, and there being in general much company, added to the difficulty of choosing the most airy rooms for his retirement.

July 8*th.*—I had a dawn of hope that my brother might regain once more a little strength, for I have a memorandum in my almanac of his walking with a firmer step than usual above three or four times the distance from the dwelling-house to the library, in order to gather and eat raspberries, in his garden, with me. But I never saw the like again.

The latter end of July I was seized by a bilious fever, and I could for several days only rise for a few hours to go to my brother about the time he was used to see me. But one day I was entirely confined to my bed, which alarmed Lady Herschel and the family on my brother's account. Miss Baldwin* called and found me in despair about my own confused affairs, which I never had had time to bring

---

* A younger sister of Mrs. Beckwith, niece of Lady Herschel.

into any order. The next day she brought my nephew to
me, who promised to fulfil all my wishes which I should
have expressed on paper; he begged me not to exert myself
for his father's sake, of whom he believed it would be the
immediate death if anything should happen to me. . . . .*
Of my dear nephew's advice I could not avail myself, for I
knew that at that time he had weighty concerns on his mind.
And, besides, my whole life almost has passed away in the
delusion that next to my eldest brother, none but Dietrich
was capable of giving me advice where to leave my few relics,
consisting of a few books and my sweeper. And for the
last twenty years I kept to the resolution of never opening
my lips to my dear brother William about worldly or
serious concerns, let me be ever so much at a loss for
knowing right from wrong. And so it has happened that at
the time when I was stupefied by grief at seeing the death
of my dear brother, I gave myself, with all I was worth, up
to my brother Dietrich and his family, and from that time
till the death of D. I found great difficulty to remain
mistress of my own actions and opinions. In respect to the
latter we never could agree. And this it was which
prompted me to send Flamsteed's works to Göttingen (I
would rather have kept them till now) for fear they might be
offered for sale. Having about this time received very distressing accounts of family misfortunes from my brother at
Hanover, I could find no rest on his account till I should
have made my £500 stock over to him, but this required my
presence at the bank, and I could not think of leaving
Slough till my brother should be engaged for some days
with his family previous to the departure of my nephew,
who was going to accompany a friend abroad. And besides,
I knew that my absence would then be scarcely perceived,

* This passage is a later note, added Sept. 26, 1828.

as a very sensible elderly lady (Mrs. Monson) would be there on a visit.

Aug. 8*th.*—I went, and at six o'clock in the afternoon of the 10th I was home again. My nephew had left Slough the same morning.

I found my brother seated by the ladies, but so languid that I thought it necessary to take a seemingly unconcerned leave for the night.

Aug. 11*th,* 12*th,* 13*th,* and 14*th* I went as usual to spend some hours of the forenoon with my brother.

Aug. 15*th.*—I hastened to the spot where I was wont to find him with the newspaper which I was to read to him. But instead I found Mrs. Morson, Miss Baldwin, and Mr. Bulman, from Leeds, the grandson of my brother's earliest acquaintance in this country. I was informed my brother had been obliged to return to his room, whither I flew immediately. Lady H. and the housekeeper were with him, administering everything which could be thought of for supporting him. I found him much irritated at not being able to grant Mr. Bulman's request for some token of remembrance for his father. As soon as he saw me, I was sent to the library to fetch one of his last papers and a plate of the forty-feet telescope. But for the universe I could not have looked twice at what I had snatched from the shelf, and when he faintly asked if the breaking up of the Milky Way was in it, I said "Yes," and he looked content. I cannot help remembering this circumstance, it was the last time I was sent to the library on such an occasion. That the anxious care for his papers and workrooms never ended but with his life was proved by his frequent whispered inquiries if they were locked and the key safe, of which I took care to assure him that they were, and the key in Lady Herschel's hands.

After half an hour's vain attempt to support himself,

my brother was obliged to consent to be put to bed, leaving no hope ever to see him rise again. For ten days and nights we remained in the most heartrending situation till the 25th of August, when not one comfort was left to me but that of retiring to the chamber of death, there to ruminate without interruption on my isolated situation. Of this last solace I was robbed on the 7th September, when the dear remains were consigned to the grave.

Sept. 9th.—I returned to my house and began selecting the books and clothing I should want to take with me to Hanover, where I thought it best to go with the Michaelmas messenger.

Sept. 27th.—I had disposed of my furniture, partly by presents and partly by sale; and after settling with my landlord, &c., I left my house for Lady Herschel's, to remain there till business should call her and my nephew to town.

Oct. 3rd.—My friends as well as myself were made easy by the arrival of my brother Dietrich, who came to fetch me.

Oct. 7th.—I took leave of Princess Augusta and all my friends and connections in Windsor.

Oct. 10th.—At 9 in the morning I left Slough with my brother D. Lady H. and my nephew followed the next day.

*   *   *   *   *

Oct. 14th.—Princess Sophia Mathilda sent her carriage for me to spend the day with her at Blackheath.

Oct. 16th.—I went with my brother to Mortlake to take leave of Baron Best and family; and thence we directly proceeded to Bedford Place, where all my friends were assembled, among whom I had the comfort of seeing once more my nephew's friend, and the favourite of my dear departed brother, Mr. Babbage. He had only that day arrived from the North. I could find no opportunity for any conversation with him, but just by a pressure of the hand recommended my nephew in incoherent whispers to the continuance of his regards and friendship.

From all these sorrowing friends and connections I was obliged to take an everlasting leave, and in the few hours we were for the last time together, I was obliged to sign many papers, among which was a receipt for a half year's legacy I signed this with great reluctance . . . . but Lady H. and my nephew insisted on my taking it, according to my brother's will. This unexpected sum has enabled me to furnish myself with many conveniences on my arrival here, of which otherwise I should have perhaps debarred myself.

Oct. 17th.—In the morning we left our lodging for an inn near the Tower. Mr. Beckwith joined us, and settled at the Custom House for our baggage. My nephew came for a moment to us, and after his departure I saw no one I knew or who cared for me.

Oct. 18th.—At ten o'clock we went on board of the steam packet.

Oct. 20th.—At noon we landed after a stormy passage at Rotterdam.

Oct. 21st.—At daybreak we began to proceed on our way, and

Oct. 28th.—We arrived at the habitation of my brother, in Hanover.

A note, dated September 29th, 1828, apologizes to her nephew for troubling him with the above and other papers, adding :—

I have destroyed my Day-book, but in doing so I was tempted to extract some dates which I thought might still be interesting to me, and bring the past once more to my recollection ; but as that would only be a drawback to the satisfaction I almost daily may enjoy by hearing of the fame of my dear nephew, it is best to remove all that can bring the past to my recollection.

The letters which follow are the only documents from which any particulars can be drawn for this and many following years. No Day-book or note-book of any kind appears to have been kept, or at any rate preserved, from the time of the return to Hanover in October, 1822, until the year 1833.

CHAPTER V.

RETROSPECTION.

As we close the record of Miss Herschel's residence in England, we may pause for a moment to look back over the space she had traversed while following, with unvarying diligence and humility, the path her brother marked out for her, first in blessed hourly companionship, when she was as necessary in his home as in his library, or among his instruments; and latterly, when with saddened heart but unflagging determination she continued to work for him, but saw his domestic happiness pass into other keeping.

While they toiled together through those first ten years of ever-deepening interest and marvellous activity, during which the rapid juxtaposition of mirror-grinding, concerts, oratorios, music lessons,* and frequent papers written for philosophical societies, almost takes the breath away as we read,—the brother had "abundant opportunity of learning how far he could trust to his companion's readiness, as well as capability,

---

* At this time W. Herschel frequently gave thirty-five and thirty-eight lessons a week to lady pupils.

to accept of duties as utterly remote from all that her previous life had prepared her for as if he had asked her to accompany him on a pilgrimage to Mecca. And thus, of all of whom he had made trial, it was not the brilliant Jacob, nor the gifted Alexander, but the little quiet, home-bred Caroline, of whom nothing had been expected but to be up early and to do the work of the house, and to devote her leisure to knitting and sewing, in whom he found that *steady devotion to a fixed purpose* which he felt it was possible to link with his own. "I did nothing for my brother," she said, "but what a well-trained puppy-dog would have done : that is to say, I did what he commanded me. I was a mere tool which *he* had the trouble of sharpening." Such was always her own modest self-estimate. It is hardly too much to say that, to have worked as she had worked, and to have done all that she had accomplished, and to claim no more than the credit due to passive obedience to orders, is a depth of humility of that rare and noble kind which is in itself a form of greatness. It must not be forgotten, that the progress of astronomical science since Sir William Herschel's great reflector startled the world, has not been greater than has been the change, both in opinion and practice, on the subject of female employments and education. The appointment of a young woman as an assistant astronomer, with a regular salary for her

services, was an unprecedented occurrence in England.
She had watched and shared in every effort and
every failure from the first seven-foot telescope to
the construction of the ponderous machinery that
was to support the mighty tube of which she herself
made the first crude model in pasteboard. When,
finally, her brother was summoned to the King, and
wrote to tell her how he fared at Court, she accepted
the decision, by which he exchanged a handsome income for the sake of obtaining the command of his own
time, and £200 a-year from his gracious sovereign,
with only a passing expression of regret from the
housekeeper's point of view, and threw herself heart
and soul into the new life at Datchet. One all-sufficing
reward sweetened her labours—" I had the comfort to
see that my brother was satisfied with my endeavours
in assisting him." When the dignity of original
discovery gave her a distinct and separate claim to
the respect of the astronomical world, she must
have found out that she was something better
than a mere tool. The requisite knowledge of
algebra and mathematical formulæ for calculations and
reductions she had to gather when and how she could :
chiefly at breakfast, and at any odd moments when
her brother could be asked questions, and the answers
were carefully entered in her Commonplace Book,
where examples of taking equal altitudes, and how to
convert sidereal time into mean time, follow upon

pages of problems, oblique plain triangles, right-angled spherical triangles, how to find the logarithm of a number given, and theorems for making tables of motion. With this slender store of attainment she accomplished a vast amount of valuable work, besides the regular duties of assistant to so indefatigable an observer as Sir William Herschel. He was invariably accustomed to carry on his telescopic observations till daybreak, circumstances permitting, without any regard to season; it was the business of his assistant to note the clocks and to write down the observations from his dictation as they were made. Subsequently she assisted in the laborious numerical calculations and reductions, so that it was only during his absences from home, or when any other interruption of his regular course of observation occurred, that she was able to devote herself to the Newtonian sweeper, which she used to such good purpose. Besides the eight comets by her discovered, she detected several remarkable nebulæ and clusters of stars previously unnoticed, especially the superb nebula known as No. 1, Class V., in Sir William Herschel's Catalogue. Long practice taught her to make light of her work. "An observer at your twenty-foot when sweeping," she wrote many years after, "wants nothing but a being who *can* and *will* execute his commands with the quickness of lightning; for you will have seen that in many sweeps six or twice six objects have been secured and described in one minute of time."

The ten years from 1788 to 1798, although a blank as regards her personal history — the Recollections cease with her brother's marriage—were among the busiest of her life, and in the year last mentioned the Royal Society published two of her works, namely, " A Catalogue of 860 Stars observed by Flamsteed, but not included in the British Catalogue," and "A General Index of Reference to every Observation of every Star in the above-mentioned British Catalogue." It is in reference to these that she wrote the very interesting letter to the Astronomer Royal, which is given among others, in its place, in the Journal. But another work, which was not published, was the most valuable, as it was the most laborious of all her undertakings. This was " The Reduction and Arrangement in the form of a Catalogue, in Zones, of all the Star-clusters and Nebulæ observed by Sir W. Herschel in his Sweeps." It supplied the needful data for Sir John Herschel when he undertook the review of the nebulæ of the northern hemisphere ; and it was for this that the Gold Medal of the Royal Astronomical Society was voted to her in 1828, followed by the extraordinary distinction of an Honorary Membership. This Catalogue was not completed until after her return to Hanover, and Sir David Brewster wrote of it as " a work of immense labour," and "an extraordinary monument of the unextinguished ardour of a lady of seventy-five in the cause of abstract science."

Although the Recollections cease in 1788, there are some volumes recording the nature and results of her nightly "sweepings," which Miss Herschel kept very regularly, and, as an unique example of a lady's journal, a few of the entries may be of interest.

1788. *Sept. 9th.*—My brother showed me the five satellites of Saturn. He made me take notice of a star, which made a double star last night with the fifth satellite.

*   *   *   *   *

Dec. 8th.—I swept for a comet which was announced in the papers as having been discovered the 26th of November by Mr. Messier. According to the observations of that date, it should have been within a few degrees of the Pole star (by my brother's calculation), but though I swept with great attention a space of at least ten or twelve degrees all around the pole over repeatedly, I could find nothing.

Another night of unavailing search, with thermometer 20°.*

1790. *Jan. 7th.*—I have swept all this evening for my [third] comet in vain. My brother showed me the G. Sidus in the twenty-foot telescope, and I saw both its satellites very plainly.

1791. *Aug. 2nd.*—I began to sweep at 1.30, from the

---

* It was not an unknown circumstance for the ink to freeze while she was attending to take down her brother's observations.

horizon through the Pleiades up as high as the head of
Medusa. Left off with β Tauri. Afterwards I continued
with horizontal sweeps till daylight was too strong for seeing any longer.

1792. *May 3rd.*—My brother having desired me by way
of practice to settle the stars α Persei and Castor, and α Virginis, by some neighbouring stars in Wollaston's Catalogue,
I made last night an attempt to take their places. The
moon was near the full, therefore no sweeping could be
done.

1795. *May 1st.*—*Mem.* In the future when any great
chasms appear in my journals, it may be understood that
sweeping for comets has not been neglected at every opportunity which did offer itself. But as I always do
sweep according to the precept my brother has given me,
and as I often am in want of time, I think it is very im-
material if the places where I have seen nothing are noted
down.

Nov. 7th.—0.40 sidereal time. About an hour ago I
saw the comet [seventh] which is marked in the annexed
field of view [diagrams drawn with extreme neatness illus-
trate the entries when necessary]. When I perceived it
first the two small stars were entirely covered by it, and it
appeared to be a cluster of stars mixed with nebulosity ;
but not knowing of such an object in that place, I kept
watching it, and perceived it to be a comet by its having
moved from the two small stars, so as to leave them entirely free from haziness.

1797. *Aug. 14th.*—C. H.'s comet. At 9.30 common
time, being dark enough for sweeping, I began in the usual
manner with looking over the heavens with the naked eye,
and immediately saw a comet nearly as bright as that which
was discovered by Mr. Gregory, January 8, 1793. I went
down from the observatory to call my brother Alexander,

L 2

that he might assist me at the clock. In my way into the
garden I was met and detained by Lord S. and another
gentleman, who came to see my brother and his telescopes.
By way of preventing too long an interruption, I told the
gentlemen that I had just found a comet, and wanted to
settle its place. I pointed it out to them, and after having
seen it they took their leave.

These entries were continued with great regularity
to the year 1819, at which time, as the Diary shows,
Sir William's increasing feebleness made her close
daily attendance more necessary, and her pen was in
greater request than the " sweeper." The last volume
concludes with a carefully drawn eye-draft of the
situation of a comet visible at Hanover, January 31st,
1824. Thenceforth the instrument which had done
such good service in her hands for forty years of
steady work, became the chief ornament of her sitting-
room, until her disquieting fears for its ultimate fate
led her to send it back to England.

Sad as is the story of those last years of declining
old age, while the beloved brother lived we know that
his sister's life was full of occupation. It is not until
the cruel hour comes, and she knows that death and the
grave will soon claim him, that she allows the sense of
her own bitter desolation to find expression. When all
was over, her only desire seems to have been to hurry
away. Hardly was he laid in his grave than she collected the few things she cared to keep, and left for
ever the country where she had spent fifty years of her

life, living and toiling for him and him only. "If I
should leave off making memorandums of such events
as affect, or are interesting to me, I should feel like—
what I am, namely, a person that has nothing more
to do in this world." Mournful words : doubly mournful when we know that the writer had nearly half an
ordinary lifetime still between her and that grave which
she made haste to prepare, in the hope that her course
was nearly run. Who can think of her, at the age of
seventy-two, heart-broken and desolate, going back to
the home of her youth in the fond expectation of finding consolation, without a pang of sympathetic pity ?
She found everything changed. In addition to those
changes, for which she might have been in some measure
prepared, there were others of a kind to admit of neither
cure nor alleviation. The life she had led for fifty
years had removed her, she little guessed how much,
from the old familiar paths : her thoughts, her habits,
all her ideas had been formed and moulded in a totally
different world : more bitter still, she found herself
alone in her great sorrow and quenchless love ; pride
in the distinction reflected on themselves from rela-
tionship to the illustrious astronomer was a miserable substitute for the reverential affection she had
looked to find for one of the kindest and most generous
of brothers. But the bitterest suffering of all was
from a source which was, and ever remained, beyond
the reach of help. " You don't know," wrote one of

Miss Edgeworth's sisters, "the blank of life after having lived within the radiance of genius;" and this was the blank in which Miss Herschel doomed herself not only to live, but to try to begin anew, when past three score and ten. The extracts from her letters bear strong testimony to the gallant struggle she made to find interests and occupations in what those about her, as well as she herself, looked upon as a kind of exile, and "Why did I leave happy England?" was often her cry, more especially as time went on, and interest in her nephew and his family came mercifully to fill the heart still so yearning and ready for affection. When she heard the news of Sir John Herschel's intended departure for the Cape, she wrote, "Ja! if I was thirty or forty years junger and could go too? in Gottes nahmen!" her interest in the science to which she had devoted her best years never ceased, though she persisted to the end in ridiculing the bare suggestion that the Rosse telescope could by any possibility be so good as *the* forty-foot. The homage paid to her as a *savante* amused as well as gratified her. "You must give me leave to send you any publication you can think of," she wrote to her nephew, "without mentioning anything about paying for them. For it is necessary I should every now and then lay out a little of my spare cash in that for the sake of supporting the reputation of being a learned lady (there is for you!), for I am not only looked at for such a one, but even stared at here in Hanover!" Her deprecation of the membership of the Irish Academy, conferred on one who for so many years had " *not even discovered a comet,*" was thoroughly sincere as well as characteristic, but she found pleasure in receiving the homage which was naturally paid to her; no man of any scientific eminence passed through Hanover without visiting her;* and it became a matter of public concern to note the presence of the well-known tiny figure at the Theatre, where her constant appearance in extreme old age was in itself a marvel. The frugal simplicity of her habits made it a positive perplexity to dispose of her income; she protested that £50 a-year was all she could manage to spend on herself, and she pertinaciously resisted receiving the pension of £100 per annum left to her by her brother, often devoting the quarterly or half-yearly payment to the purchase of some handsome present for her nephew or niece. She wrote full instructions and made the most careful arrangements for every detail of business in connection with her own burial and the disposal of her property—that is of the little she reserved, for her generosity towards her relations was as great as the expenditure on herself was small.

In these last remarks I have anticipated events, and

---

* From the Royal Family she received the most kind and graceful attentions.

must now return to the year 1822, when the correspondence begins.

ROTTERDAM, *Monday, Oct.* 21, 1822.

DEAR LADY HERSCHEL,—

At this present moment I have nothing to wish for,
besides the means of convincing myself by one look of your
and my dear nephew's health. After a very troublesome
passage of forty-eight hours, we find ourselves almost restored to our former condition and composure, with only the
difference that we have no more hunting after our trunks
from Custom-house to Custom-house, and can proceed on
our way to Hanover in peace after one night's rest here in
a very good inn. But the last night was truly dismal, for
the sailors themselves confessed that it was what is called a
high sea. At one time a spray conveyed a bucket-full of
water into my bed, which was regarded as nothing in comparison to the evils with which I was surrounded. I was
the most sick of all on board, and the poor old lady was
pitied by all who enquired after her, but I had four ladies
in the same cabin with me, who encouraged me to hold out,
which at one time I thought would have been impossible.
Something happened to the vessel for want of a good pilot
in the Thames, and at Blackwall we laid still three hours,
then we hobbled on to near Gravesend, and there lay in a
high sea at anchor all night, whilst they were hatching
and thumping to mend the vessel we were to go in. In
consequence of this, we could not reach the spot where a
pilot could meet us time enough on Sunday evening, and
lay again at anchor. At half past eleven I set foot on
shore, where so many people were assembled to gaze on us
that it set me a crying, and now I am glad to be shut up
once more in a room by myself and where I can make proper

preparations for travelling further, which hitherto I have not had the opportunity of doing. All my clothes which I had prepared for the ship or sleeping on the road were locked up at the Custom-house, and I could not get hold of them again till we entered this house. So much for our adventures at present, and I beg and hope you will soon and often let us know how you are with my nephew, and how and where you can pass the following winter months in the most comfortable way.

My brother is gone into the street to look about him. The weather is fine, and I wish my dear nephew was with him, for it looks very tempting and new all about me, and I think he would enjoy seeing the bustle on the water with which this house is surrounded. My brother has charged me with millions of compliments and thanks to yourself and our nephew, but I cannot afford him quite so many, as else there would be no room for all those I owe to my dear Lady H. and my nephew, who took last Friday so long a walk to see us once more. My fears for what was to come and regret for what I left behind were so stupifying that it made me almost insensible to all what was passing about me, only this I shall remember, with satisfaction, that his looks were better than I have seen for a long time past.

I am now going to direct the little parcel for Professor Swinden, and likewise to Mr. Crommelin, jun., and to Professor Moll, at Utrecht, and Gauss will not be forgotten as we go along.

I beg you will remember me to Miss Baldwin (who I hope is with you), and particularly to Mr. Beckwith, whom I shall never be able to thank sufficiently for the friendly care he has shown to me on all, and especially on the last occasion of helping me on with my packages.

Farewell, my dear Lady Herschel, and let me hear soon that you and my nephew are well.

Miss Baldwin will write, and of course she will inform me of her own and all friends' health, &c.

Ever your affectionate
CAR. HERSCHEL.

FROM MISS HERSCHEL TO LADY HERSCHEL.

HANOVER, *Oct.* 30, 1822.

MY DEAR LADY HERSCHEL,—

We arrived here at noon, on the 28th, without the least accident, but not without the utmost exertion and extreme fatigue to both my brother and myself, from which it will be some time before I shall get the better, on account of the many visits of our friends, who come to convince themselves of our safe arrival, of which I hope you will have been informed long before this can reach you, as Mr. Quintain has promised me to send you a line the moment he reaches London. He left Hanover yesterday. I had wrote a letter in hopes he would have taken it, but that was impossible, and the post from here has been changed from Tuesday to Monday.

Mr. Housman called also here yesterday, and you may easily imagine that many inquiries are made after you and my dear nephew by all those who come near me, and I hope you will soon enable me, by a few lines, to inform them of your welfare and health, and give me the comfort to know that you have regained some of your former composure, after the late melancholy change and unsettled state in which we all were involved.

I found Mrs. H. in personal appearance so different from what I had imagined, that I can hardly believe her to be the same; she is just sixty-three years of age, and suffers much from rheumatism, which has taken away partially the use of her hands, but she is still of so cheerful a disposition and so active by way of overcoming disease by exercise, that I cannot wonder enough, and her reception of me was truly gratifying; the handsomest rooms, three or four times larger than what I have been used to, from which I can step in her own apartments, have been prepared for me and furnished in the most elegant style. But I cannot say that I feel well enough to enjoy all these good things nor be able to show myself to those who wish to see me, at least not at present.

Mrs. Beckedorff sent to enquire after me when I had been hardly two hours arrived. Miss B. is confined with a severe cold. My brother went yesterday to see them, and we have postponed our meeting till Saturday, when she will come to town for the winter.

From Rotterdam I sent a letter which I hope you have received, and by which you will have seen that our passage was not of the most agreeable kind.

The papers to Professor Van Swinden, Crommelin jun., at Amsterdam, and Professor Moll, at Utrecht, have been delivered, but that to Gauss, I am sorry to say, is either lost or mislaid, for I cannot find it anywhere, and I am vexed to give my dear nephew so bad a sample of my willingness to be of use to him. Perhaps through Mr. Quaintain he might get one over when the Duke of Cambridge returns, else the next conveyance I know of is at Christmas, by Gotterman.

I beg my love to my nephew and Miss Baldwin, who, I hope, will soon let me know how you are, &c.

Believe me,
Your truly and affectionate
CAR. HERSCHEL.

FROM MISS HERSCHEL TO LADY HERSCHEL.

HANOVER, *Nov.* 12, 1822.

MY DEAR LADY HERSCHEL,—

I hope you have received the letter which I sent by
the first post which went from here after my arrival, dated
31st October, and also one I wrote in Rotterdam, by which
you will have seen what a disagreeable passage we had at
sea, but all those frights and fears, and the troubles and
fatigues of the journey we afterwards experienced by land
appear now to have been nothing but a dream, and my
waking thoughts are for ever wandering back to the scenes
of sorrows which embittered the afflicting and final parting
from my revered brother. If I could but be assured that
you and my dear nephew at this present moment were in
tolerable health and otherwise exempted from vexation, I
should feel myself much more comfortable, but it is hard to
live for months without knowing what may have happened
to those with whom one has been for so many years immediately connected and in the habit of keeping up a daily
intercourse.

I have hitherto not been able to overcome a dislike to
going abroad, and what little I have seen of Hanover (in my
way to the families of my two nieces and Mrs. Beckedorff
who live all close by) I do not like! And though some
streets have been enlarged (as I am told), they appear to me
much less than I left them fifty years ago. But a total
seclusion from society will not do for a continuance, for I
will not be ungrateful, I must call on the Delmerings, &c.,—
who have been here. Mrs. D. is grown quite fat and very
handsome, her daughter is a head taller and a very pretty
young woman; the eldest son is already in the service with
the Erz Herzog of Strelitz, and there has been no increase
in the family since they left England. Mrs. D. made many

inquiries after you and my nephew's health, and gratefully remembers the kindly treatment she received at all times from you.

Nov. 18*th.*—Mrs. Beckedorff and Miss B. and myself have been laid up with severe colds, and I am still unable to go into company, but Mrs. B. sent Dr. Mury to make her excuse for not returning my visit. The first time I went to them, Mrs. B. made all her ten grandchildren stand up before me according to their ages, and a fine healthy family it is. But all the little folks I am introduced to are disappointed at finding me to be only a *little* old woman; which I suppose must be owing to having been told the *Great* Aunt Caroline from England was coming.

From the family of my eldest niece I have seen nothing as yet, and probably shall not before next summer, as her affairs must remain for some time in an unsettled state. I did not know till we were within sight of Hanover how greatly I was obliged to my brother for coming to fetch me, for I find he was but barely recovered from a serious illness when he left home, which had been occasioned by travelling to and fro to his daughter, who was in need of the support of both her parents on losing her husband after a few days' illness; in the same week she had given birth to a son, and was made a widow with nine children in her 38th year. But, happily, she is blessed with an uncommon share of understanding and fortitude, besides the means of seeing them well educated and improving their fortunes.

Nov. 27*th.*—You will see, my dear Lady H., by the above, that at different times I have been employed in giving a circumstantial account of all what concerns that part of my family amongst whom I came to end my days; but I would not conclude, nor send off my letter, till I should have received some satisfactory account of your well being, and the arrival of the last post has given a most agreeable turn to the dismal impression the parting scenes of the 17th and 18th October had left on my mind. To Miss Baldwin I feel greatly obliged for her comforting letter, and hope she will be able to write me many more equally consoling; my brother is going to speak for himself, and if I would leave a little room for a few words to my nephew, I must conclude with saying that I am

My dear Lady Herschel's

Most obliged and affectionate,

Car. Herschel.

My dear Nephew,—I thank you for the few lines in the P.S., for by them I see you were thinking of me when you procured some indexes to Flamsteed's obs. But I will not trouble you to send any; I only wished you to have some for your own friends, Mr. South, Major Kater, &c., for as they were not members of the R. Society at the time of publication, they may perhaps not be possessed of that necessary Appendix.

The next messenger will take the book Mr. Babbage wishes for, and I want very much to send you some of the numerous philosophical productions in which this country my nephew Grosekopf says abounds, but I am at a loss on what to fix my choice. *I wish you would let me know if any of the works of Schelling are known in England?* Of him it is said that his philosophy is entirely new, and beyond all what goes before, and so profound, that nobody *here* can understand him, &c.

Believe me yours most affectionately,

Car. Herschel.

FROM MISS HERSCHEL TO LADY HERSCHEL.

HANOVER, *Dec.* 18, 1822.

My dear Lady Herschel,—

At last I am enabled to inform you of the safe arrival of my boxes and trunks, which only came the day before yesterday, and then I was obliged to wait till the keys were sent by to-day's post, but I have the satisfaction to find that every article is exactly as I had packed them with my own hands. For the last three weeks, I was despairing of ever seeing them again, for the vessel had been no less than three weeks at sea, and then had been obliged to unload six German miles beyond Bremen for want of water in the Weser. The country is in general much distressed for want of water; our large rivers may be passed on foot, &c. But of these things you are perhaps informed by the newspapers, and of many other circumstances; such as the mice eating the corn as soon as sowed, so that sowing it three times over was without effect, till the mice were destroyed by a pest coming among them.

I would give anything if I at this moment could see with my own eyes how you and my dear nephew are; tell him that on the day after Christmas (Dec. 26th) the messenger will leave Hanover, and will take the book for Mr. Babbage, and one in two volumes for my nephew; also two or three letters of his father's which I have found among some papers of my brother Alex.

I know not if I mentioned it in my last that I selected all his last receipts when he left England, and shall keep them yet a little longer.

As yet I lead but a dull sort of life; the town is much too gay for me—plays, concerts, card parties, walking, &c. I cannot take part *in any;* my cold in my head is still very bad, and my poor brother is frequently unwell, and for want of my trunks I could not accept Mrs. Beckedorff's invitation to meet Madam Zimmermann, &c., in an evening, on account of not having clean things; but she is so kind as to call on me sometimes among all the hurry she is engaged in at present with the Princess Augusta.

Mr. Gisewell came a few days ago to see me; he lives a little way out of town, and poor Mrs. G. keeps her bed, and is hardly ever well; their eldest daughter is happily situated with the Queen of Würtemberg, and Mr. Gisewell enjoys a very lucrative situation.

I wish you could conveniently acquaint my nephew, H. Griesbach, as soon as possible, that my brother has received an answer to the letter he sent to Antwerp to the sister of H. Griesbach, and that in the parcel which the messenger will bring will be enclosed her letter to Mr. H. G.

I hope you will make Miss Baldwin write me soon a long account how yourself and all *around* and *with* you are, but pray let it be a favourable one, and remember me to all (who are so good as to inquire after me) most cordially, and believe me,

My dear Lady H.,
Your very affectionate
CAR. HERSCHEL.

P.S.—We have had a few days' very severe frost; tomorrow I shall unpack my thermometer; I suppose I shall find the difference between a German and an English winter, though they make the rooms hot enough with their stoves; but then I am afraid of firing their chimneys, and we have no water, though the police have demanded that every housekeeper shall be provided with eight buckets of water in their kitchen; besides, the price of fuel is enormous, owing to the French having destroyed all the forests.

FROM MISS HERSCHEL TO J. F. W. HERSCHEL, ESQ.

HANOVER, *Dec.* 26, 1822.

MY DEAR NEPHEW,—

The parcel I am packing up contains so many odds and ends, that I think it will be necessary to give you an inventory of them. The most interesting to you, I think, will be the three letters from your dear father (which I found among my brother Alexander's papers), both on account of the handwriting and their containing some accounts of the busy life of the times in which they were written.

Of the philosophical work, I will say nothing further than that I am curious to know if I have sent you sense, or nonsense, that I may know in future how to trust my informer; I am only sorry I could not send them bound, but they came too late from Leipsic for that purpose. In the small cover (with *your* little man looking through the telescope) is a shade of your Uncle Alex., which you will be so good as to give to your mother, who (if I remember right) wished for the same, after it had been packed up, and she will perhaps be so good as to send the letter to Mr. Henry Griesbach the first time anybody goes to Windsor.

So much for business, and on the other side I will talk a little of myself. But it is a poor account I can give of myself at present, and the worst of it is that I cannot hope for better times. I am still unsettled, and cannot get my books and papers in any order, for it is always noon before I am well enough to do anything, and then visitors run away with the rest of the day till the dinner hour (which is two o'clock). Two or three evenings in each week are spoiled by company. And at the heavens is no getting, for the high roofs of the opposite houses.

But within my room I am determined nothing shall be wanting that can please my eye. Exactly facing me is a bookcase placed on a bureau, to which I will have some

M

glass doors made, so that I can see my books. Opposite this, on a sofa, I am seated, with a sofa-table and my new writing-desk before me, but what good I shall do there the future must tell.

Many more of such like transactions I was going to communicate to you, but I am interrupted by the carpenter (our Andrews), who is come to do some jobs for me, so for this once you will be released from my nonsense.

But one thing I must yet add, which is that you will accept my heartfelt wishes for your health, happiness, and prosperity throughout the coming year and for many more hereafter, in which my brother and sister are joining most sincerely, to yourself and Lady Herschel, and believe me, my dear nephew,

Ever your most affectionate aunt,

CAR. HERSCHEL.

FROM MISS HERSCHEL TO J. F. W. HERSCHEL, ESQ.

HANOVER, *Feb.* 27, 1823.

MY DEAREST NEPHEW,—

I take the earliest opportunity I have to acquaint you with having received a letter from Mr. H. Goltermann, accompanied with a draft for £2 4s. 6d., which is already received and safely deposited in my writing-desk. But the information that he had had the pleasure of seeing you in good health afforded me the *greatest* satisfaction, and he further promised me to forward the parcel to you in Downing Street, which was particularly pleasing to me, as I wished to avoid the sending backward and forward by blundering coachmen.

On the 5th of this month I received your letter without date, but conclude it was written about the same time with those of your dear mother and cousin Mary, dated the 9th and fifteenth of January. I delayed answering them (and must do so still for the present) because I knew that all mails were detained this side of the sea.

One passage in your letter affected me much, it was gratifying to me and unexpected : ". . . . speaks of your *English life*, &c. . . But now that you have left the scene of your labours you have the satisfaction of knowing that they are duly appreciated by those you leave behind." But I can hardly hope that those favourable impressions should be lasting, or rather not be effaced by my hasty departure ; but believe me I would not have gone without at least having made the offer of my service for some time longer to you, my dear nephew, had I not felt that it would be in vain to struggle any longer against age and infirmity, and though I had no expectation that the change from the pure country air in which I had lived the best part of my life, to that of the closest part of my native city, would be beneficial to my health and happiness, I preferred it to remaining where I should have had to bewail my inability of making myself useful any longer.

I hope you and Lady H. have not suffered by the severity of the weather ; to me it has certainly done no good. I am grown much thinner than I was six months ago ; when I look at my hands they put me so in mind of what your dear father's were, when I saw them tremble under my eyes, as we latterly played at backgammon together. Good night ! dear nephew, I will say the rest to-morrow.

By way of postscript I only beg you will give my love and many thanks to your dear mother and cousin for their kind letters ; and if the latter will continue from time to time to inform me of all your well-being, I shall equally feel gratified, for it is no matter from which hand I receive the comfortable information.

I remain, ever your affectionate aunt,

CAR. HERSCHEL.

M 2

FROM MISS HERSCHEL TO J. F. W. HERSCHEL, ESQ.

HANOVER, *April* 14, 1823.

MY DEAREST NEPHEW,—

I hasten to send this sheet, which is but this moment come to hand, and the post within an hour of leaving Hanover. I begin to fear that I shall not hear from you till you send me an acknowledgment of having received the certificate, which we are not able to obtain till after the 10th of April and 10th of October, but January and July it is the 5th. I assure you I would rather go without the money than be so long without hearing from you, or have a line to express your pleasure for the present I offered you and Mr. Babbage by sending the books by the Christmas messenger, of which I, at this moment, have no information that they have been delivered. By the Easter messenger I have sent some metwurst [a Hanoverian delicacy], which I hope you and your dear mother will find good, but when they are once cut they must be eaten soon, else they are dry and lose all their flavour.

*   *   *   *   *

The Germans are very busy about the fame of your dear father; there does not pass a month but something appears in print, and Dr. Groskopf saw in den gelehrten Zeitungen that Professor Pfaff had translated *all* your dear father's papers from the Phil. Trans. into German, and which will be published in Dresden. I wish he had left it for some good astronomer to do the same. Pray let me know how you and your dear mother are in health; I am not well, but have a severe cold at present, but am always and still your affectionate aunt,

CAR. HERSCHEL.

The following letter from the Princess Sophia of Gloucester is a pleasing memorial of the kindness and amiability of which Miss Herschel experienced so many proofs while she lived at Slough :—

THE PRINCESS SOPHIA MATILDA TO MISS HERSCHEL.

MY DEAR MISS HERSCHEL,—
    Your obliging attention in sending the Astronomical Almanack to me I am very sensible of, and at the same time that I return my best thanks for this flattering mark of your recollection, I must express my regret that I am not possessed of more knowledge and leisure, that I might profit sufficiently by your kindness in endeavouring to instruct me. I was very happy to learn that you had reached your native land in safety, and I sincerely form every wish that your health may be *long* preserved to you!
    May I request you to remember me kindly to Mr. and to Miss Beckedorff, and to be assured yourself of the true esteem and regard with which I remain, my dear Miss Herschel,
                    Yours very faithfully,
                            SOPHIA MATILDA.
LONDON, *June* 16, 1823.

FROM MISS HERSCHEL TO J. F. W. HERSCHEL, ESQ.

                            HANOVER, *June* 24, 1823.
MY DEAREST NEPHEW,—
    I had intended to write you a long and very learned epistle, but I am just now informed that the messenger will leave Hanover within a very few hours, and I must content myself with giving you the outlines of what I would have said.
    I believe I have mentioned in a former letter to your mother that a Professor Pfaff has announced his intention of giving a translation of your father's papers. It runs in

my head that this professor is but a *Jackanapes*,* who will spoil the *broth*,* and I wished he would not meddle with what he cannot understand. But I thought it but right to inform you of what is come to my knowledge, particularly as I was told it had been announced again that the translation would appear with corrections and explanations. Dr. Luthmer (in the "Ast. Jahrbuch" for this year you may see a paper by this gentleman) told me since there were two professors of that name (brothers), one an *astrologer*, and if it was the latter he would make nonsense of it.

Miss Baldwin mentioned you were at Cambridge on the business of having your father's papers printed. I think it could not be amiss if something of your intention could be mentioned in the *Edinburgh Quarterly Review*, which appears here at Hanover, and of course throughout Germany, that it may be known that your father's labours are in *yours* and of course in the *most able hands* to make remarks on them. I only wish to draw your attention this way, but say nothing.

I have mentioned it over and over again that I was so unlucky as to lose the paper on my journey you entrusted to my care for Prof. Gauss. If you have another copy to spare give it to Mr. Golterman for the return of the messenger; for he has heard of your good intention, and laments my negligence; I shall be introduced to him shortly, when he comes through Hanover again, where he passed through about a fortnight ago on a journey of observation, tending to establish some new discovery of his own, of which we are soon to know more. The theodolite has something to do with it; so much I snapt up in a company of learned ladies who, within these last two months, have taken me into their circle. But I am imitating Robinson

---

* These words had apparently to be sought for in the dictionary, as they are inserted in pencil in blank spaces left for the purpose.

Crusoe, who kept up his consequence by keeping out of sight as much as possible when he acted the governor, and when they want to know anything of me, I say I cannot tell! . . . . I did nothing for my brother but what a well-trained puppy dog would have done, that is to say, I did what he commanded me.

I send you a small publication which I think must interest you, but if it contains anything which is new to you I cannot tell. I shall, however, obtain what I very much long for, viz., to see your handwriting, for surely you will write me a line of thanks?

I am in general too unwell to sit much at the writing-table, and have not been able to do anything which could be of use to you. The letters which you will receive under cover to you I hope you will do me the favour to cause them to be safely delivered. They are sealed already, else I should have added a P.S. to your dear mother of the following, viz., that I was agreeably surprised by a letter this morning from the Princessin Sophia of Gloucester, and that my brother's family are all well at present; my brother in particular makes work for the tailor to let out his waistcoats, and they are happy to have their eldest daughter for a fortnight with them on a visit; she is a truly interesting little delicate creature just turned of forty, and has one daughter fit to be married, two sons preparing for the university, and the youngest weaned a month ago; she is to me a wonder when I look at her, she reads English fluently, French she was used to speak like her mother tongue from her infancy.

I am interrupted, and must seal up the packet.

And I remain, dear nephew,

Your most faithful and affectionate aunt,

CAR. HERSCHEL.

FROM MISS HERSCHEL TO J. F. W. HERSCHEL, ESQ.

HANOVER, *July* 14, 1823.

MY DEAREST NEPHEW,—

As a proof of my being still among the number of the living, you will perhaps not dislike to see my own handwriting added to that of the three gentlemen who signed my certificate. But I am at a loss for a subject which should be interesting to you, because, hearing so seldom from you, I begin to fear my correspondence may turn out to be troublesome. But still I long to hear a little oftener that you and your dear mother are well; for since April eleventh (date of Lady H.'s letter) I have had no assurance of the same on which I could depend.

*   *   *   *   *

I wish often that I could see what you were doing, that I might give you a caution (if necessary) not to overwork yourself like your dear father did. I long to hear that the forty-foot instrument is safely got down; your father, and Uncle A. too, have had many hair-breadth escapes from being crushed by the taking in and out of the mirror; but God preserve you, my dear nephew, says

Your most affectionate Aunt,

CAR. HERSCHEL.

P.S.—My brother and family join me in many compliments to you and your dear mother. They are all well; I am the only one who is complaining, but I think I have a right to that preference, for I am the oldest.

FROM J. F. W. HERSCHEL TO MISS HERSCHEL.

DOWNING STREET, *August* 1, 1823.

DEAR AUNT,—

I have been long threatening to send you a long letter, but have always been prevented by circumstances and want of leisure from executing my intention. The truth is, I have been so much occupied with astronomy of late, that I have had little time for anything else—the reduction of these double stars, and the necessity it has put me under of looking over the journals, reviews, &c., for information on what has already been done, and in many cases of re-casting up my father's measures, swallows up a great deal of time and labour. But I have the satisfaction of being able to state that our results in most instances confirm and establish my father's views in a remarkable manner. These inquiries have taken me off the republication of his printed papers for the present.

I think I shall be adding more to his fame by pursuing and verifying his observations than by reprinting them. But I have by no means abandoned the idea. Meanwhile I am not sorry to hear they are about to be translated into German. There is a Mr. Pfaff, a respectable mathematician, and I hope it is he who undertakes the work. If you can learn more particulars, pray send them to me. I hope this season to commence a series of observations with the twenty-foot reflector, which is now in fine order. The forty-foot is no longer capable of being used, but I shall suffer it to stand as a monument.

*   *   *   *   *

I am much obliged to you for the book on temperaments you were so kind as to send me, which seems interesting, but I have not had time to read it through. . . . .

P.S.—Your books on animal magnetism, and that for Babbage, arrived safe. . . . . I wish you would procure and send me Pfaff's translation of my father's papers as soon as published. Write as often as you can. Your letters are very interesting. I wish I were a better correspondent, but my time is so occupied, I know not where to turn.

P.P.S.—Babbage has had £1,500 granted him by Go-vernment to enable him to execute his engine, which is very curious. A report is strongly current of Captain Parry's successful arrival at Valparaiso; it comes in a very probable form.

FROM MISS HERSCHEL TO J. F. W. HERSCHEL.

HANOVER, *August* 11, 1823.

MY DEAREST NEPHEW,—

I thank you most heartily for your kind care and punctuality in sending my remittance, and am only sorry to trouble you so often; I might have acknowledged the receipt thereof by the last post, but I wished first to enable myself to give the following information. Johann Wilhelm Pfaff, professor, in Erlangen, is the same who intends to translate your father's papers, but those only which he can get a copy of. The Philosophical Transactions, I am told, are not within his reach. You may depend on my sending you whatever may come out as soon as it makes its appearance.

I can easily imagine how little time you can have to spare for writing to me when once you have entered on that mass of your father's observations contained in his journals, &c. . . . . I think the temporary index (such as it is) will in many instances be of service to you, but I wish to point out here that about the year 1800 there was a change made in the titles of some of the books. The first volume of mis-cellaneous observations was then called *Journal No.* 10, &c., ... so if the index directs you to January 24th, 1797 M. (for *M.* read *J.*) I think a memorandum of this will be found in the cover or beginning of the index, but I am not certain.

You have truly gratified me by sending the inscription of the monument,* for such subjects only are capable of inte-

---

* To her brother, in Upton Church, near Slough.

resting my waking thoughts and nightly dreams. I was going to give you an idea of what they are ; but why should I communicate grief ?

The paper for Gauss is gone to Göttingen. I have directed it to Professor Harding, who is the next to Gauss in the astronomical department, as Gauss is not yet returned from his journey of measurements. I made a few extracts from the paper * by way of having something to be delighted with, but am glad such a thing was not invented fifty years ago, for then my existence would have been of no use at all at all.

I am amusing myself with having the seven-foot mounted by Hohenbaum, though I have not even a prospect of a window for a whole constellation, but it shall stand in my room and be my monument—as the forty-foot is yours. When Hohenbaum comes for a trifling direction, we generally do not separate till dinner, or some other interruption puts a stop to our conversation; for this man is never tired when speaking of your father's inventful imaginations and the readiness with which everything was executed.

I have not above six hours' tolerable ease out of the twenty-four, and not one hour's sleep, and yet I wish to live a little longer, that I might make you a more correct catalogue of the 2,500 nebulæ, which is not even begun, but hope to be able to make it my next winter's amusement.

I was much pleased with the partial success of Mr. Babbage in having something granted towards going on with his *grand* ideas.

With many compliments and best wishes, &c.,

Your most affectionate aunt,

CAR. HERSCHEL.

* The paper referred to is probably one on "The Aberrations of Compound Lenses and Object Glasses," read at the Royal Society on the 22nd March, 1821.

CATANIA (SICILY), *July* 2, 1824.

DEAR AUNT,—

The last time I wrote to you from Slough I little expected that my next would be dated from the foot of Etna—but I mean this to be the farthest point of my wanderings, and from hence to turn my steps northwards. I am not without some hopes that my time will so far serve as to enable me to pay you a visit at Hanover, as I long very much to see you among your and my Hanoverian friends. . . . . My mother will have told you of my arrangements,—of the alteration which my plans of life have undergone (and for which I see every day more reason to be thankful), and of my present excursion, so that the date of this will not surprise you. To-morrow I hope to see the sun set from the top of Etna, and will keep this open to give you an account of my excursion there. Meanwhile let me congratulate you on the good accounts my mother gives me of your present state of health and spirits, the knowledge of which has enabled me to give real pleasure to many who, when they heard I was related to you, enquired with the greatest interest respecting you. Among the rest I may mention M. Arago, of the Observatory at Paris, and M. Fourrier, the secretary of the Institute, who has just been reading the *Eloge* of my dear father at a meeting of that body, in which I am sure (from the associations I had with him, and the written communications that passed between us on the subject) your own name will stand associated with his in a manner that cannot fail to be gratifying to you. I have not (of course, as I quitted Paris before it was read, or even written) seen it, but the man is of the right sort, and I will endeavour to procure copies of it for you and my uncle. Indeed, at Paris I find (as where do I not find it?) universal justice rendered to my father's merits, and a degree of admiration excited by the mention of his name that cannot fail to be gratifying to me, as his son. In fact, I find myself received wherever I go by all men of science, for his sake, with open arms, and I find introductions perfectly unnecessary. At Turin I sent up my card to Prof. Plana, of the Observatory, one of the most eminent mathematicians of the age, who received me like a brother, and made my stay at Turin, which I prolonged a week for the sake of his society, very pleasant. He married *a niece of Lagrange* (not of *Lalande*), and both he and his wife were full of enquiries about my "celebrated sister," (for everybody seems to think me your brother, instead of nephew), and made me tell them a thousand particulars about you. The same reception, but, if possible, still more friendly, and the same curiosity (and, I may add, the same mistake) I met with at Modena, from Professor Amici, an artist and a man of science of the first eminence. He is the only man who has, since my father, bestowed great pains on the construction of specula, and I do assure you that his ten-foot telescopes with twelve-inch mirrors are of very extraordinary perfection. Among other of your enquiring friends I should not omit the Abbé Piazzi, whom I found ill in bed at Palermo, and who is a fine respectable old man, though I am afraid not much longer for this world. He remembered you personally, having himself visited Slough.

Naples, Aug. 20th, 1824.—I take the first moment of leisure to proceed with this. I made the ascent of Etna without particular difficulty, though with excessive fatigue. The ascent from Catania is through the village of Nicolosi, about ten miles from Catania, almost every step of which is covered with the tremendous stream of lava which, in 1669, burst from the flanks of the mountain, near Nicolosi, and overwhelmed the city. Here I found a M. Gemellaro, who was so good as to make corresponding observations of the barometer and thermometer during my absence, while his brother observed below at Catania, and I carried up my mountain barometer and other instruments to the summit. From Nicolosi the ascent becomes rugged and laborious, first through a broad belt of fine oak forest, which encircles the mountain like a girdle about its middle, and affords some beautiful romantic scenery—when this is passed we soon reach the limits of vegetation, and a long desolate scorched slope, knee-deep in ashes, extends for about five miles to a little hut, where I passed the night (a glorious starlight one) with the barometer at 21·307 in.—and next morning mounted the crater by a desperate scramble up a cone of lava and ashes, about 1,000 feet high. The sunrise from this altitude, and the view of Sicily and Calabria, which is gradually disclosed, is easier conceived than described. On the highest point of the crater I was enveloped in suffocating sulphurous vapours, and was glad enough to make my observation (bar. 21·400) and get down. By this the altitude appears to be between 10 and 11,000 feet. I reached Catania the same night, almost dead with the morning's scramble and the dreadful descent of near thirty miles, where the mules (which can be used for a considerable part of the way) could scarce keep their feet.

Florence, Aug. 16th, 1824.—In the hurry and bustle of travelling one is obliged to write by snatches when one can. · · · · I hope to hear from you at all events when I reach England if I should not see you first, of which I begin now to have serious doubts, having been so terribly retarded in my Sicilian journey, and at Naples, on my return, by the illness of a friend.

Your affectionate nephew,
J. F. W. HERSCHEL.

P.S.—Have you heard how M. Pfaff's translation proceeds? I wrote to him from Cattagione, in Sicily.

MUNICH, *Sept.* 17, 1824.

My dear Aunt,—

*   *   *   *   *

I had originally intended to have gone to Switzerland from Inspruck, or from this place, having a great desire to visit the north of Switzerland, and to make certain observations among the Alps, but my wish to see you once more, to assure myself and to be able to report to my mother how I find you—to pay my uncle Dietrich a visit—and, though last, not least, to see my father's birth-place—these considerations outweigh the attractions of Switzerland, and, although the increase this détour will make in the length of my journey homewards is so considerable as to limit my stay in Hanover to two or three days at the utmost, I shall at least have had the satisfaction of not neglecting an opportunity which *may* never occur again.

The time when I hope to arrive I cannot precisely fix, as it will depend on circumstances which may occur in my route, having so arranged as to take in a variety of objects interesting in various ways, thus:—I shall go somewhat out of my way to visit Professor Pfaff, at Erlangen, and I hope also to find Mr. Encke at Seeberg, Mr. Lindenau at Gotha, Messrs. Gauss and Harding at Göttingen, &c. Moreover, I hope there will not take place a resurrection among the bones in the cave at Bayreuth before I get there. These things necessarily interrupt post haste, besides which there are always delays in passing frontiers, and accidents happening to wheels, springs, screws, &c. Allowing for these, however, I think it cannot be less than a fortnight, nor more than three weeks from the date of this when I shall have the happiness of once more shaking you by the hand, and I need not say what satisfaction it will give me to find yourself and my uncle, Mrs. Herschel and their family in good health, as well as our good friends the Beckedorffs,

Detmerings and Haussmann, with whom it will be a great pleasure to me to renew my acquaintance. You have heard, I daresay, through my mother, of our poor friend, Miss Deluc's death. Mrs. Beckedorff will have been much grieved at it.

I hope you have not forgotten your English, as I find myself not quite so fluent in this language as I expected. In fact, since leaving Italy, I have so begarbled my German with Italian that it is unintelligible both to myself and to everyone that hears it; and what is very perverse, that though when in Italy I could hardly talk Italian fit to be heard, I can now talk nothing else, and whenever I want a German word, pop comes the Italian one in its place. I made the waiter to-day stare (he being a Frenchman) by calling to him, "Wollen Sie avere la bontà den acete zu apportaren!" But this, I hope, will soon wear off.

    *   *   *   *   *

I remain, dear aunt,
Your affectionate nephew,
J. F. W. H.

FROM MISS HERSCHEL TO J. F. W. HERSCHEL.

HANOVER, *Sept.* 25, 1824.

MY DEAREST NEPHEW,—

I hardly know how to thank you sufficiently for your valuable letters, especially for the one dated the 17th of this month, as I am now at last assured that my eyes shall once more behold the continuation of your dear father. For the remaining days of my life can only by a few hours' conversation with you be made tolerable, by affording me your direction how to *finish* a general catalogue of the 2,500 nebulæ, &c., which would have otherwise caused us both a tedious and vexatious correspondence in the future.

I anxiously forbore to express my wishes for seeing you,

for fear it might have had any influence on th direction of
your intended tour.   But now all will be well, and I shall
only say that we are counting the days and hours until we
shall have the happiness of seeing you, and you will, on
entering Hanover, have only to direct your postilion to the
Markt Strasse, No. 453, where the arms of my brother and
sister, as well as mine, are longing to receive you, and till
then

> Believe me, my dearest nephew,
> Your faithful and affectionate aunt,
>   CAR. HERSCHEL.

P.S.—I beg my respects to . . Blumenbach, and I shall
ever remember with many thanks the visit with which he
honoured me when last at Hanover.

FROM MISS HERSCHEL TO LADY HERSCHEL.

HANOVER, *Oct.* 14, 1824.

MY DEAR LADY HERSCHEL,—

My dear nephew has now been gone a week, and I follow him in idea every inch he is moving farther from us, and
think he must now be near the water.   I am at this moment
in the greatest panic imaginable, for we have had all the
week much rain, and now it blows a perfect hurricane.   I
shall not send this till I have heard from you that the dear
traveller is safely at home, for it would be cruel to augment
your anxiety, which I know you are feeling till you see him
again.

[Here follows a long history of the younger members of
the Griesbach family, with details of the events of seventy
years before.]

. . . . I have not yet done, my dear Lady Herschel, and
shall not be easy till I have given some little account of my
brother's [Dietrich's] family, merely for yours and my dear

N

nephew's gratification; for, from his kind inquiries if I wanted anything? if he could do nothing for me? it seemed as if he thought he could not do enough for us. My answer was *nothing! nothing!* and this I could say with truth, as at my age and situation (which is truly respectable) I should not know what to do with more without lavishing it on others, where it would only create habits of luxury and extravagance. The time of our dear nephew's being here was too short for much confidential conversation, else I wished to have made him better acquainted with mine and my brother Dietrich's sentiments concerning the noble bequest of our lamented brother, of which Dietrich had not the most distant hope or expectation (for I believe they never had any conversation on the subject), as I am sure his way of thinking is similar to mine, that brothers and sisters (such as we were), each beginning the world with *nothing* but health and abilities for getting our bread, ought to feel shame at taking from the other if he should by uncommon exertion and perseverance have raised himself to affluence. According to this notion I refused my dear brother's proposal (at the time he resolved to enter the married state) of making me independent, and desired him to ask the king for a small salary to enable me to continue his assistant. £50 were granted to me, with which I was resolved to live without the assistance of my brother; but when nine quarters were left unpaid I was obliged to apply to him, as he had charged me not to go to anyone else. In 1803, you and my brother insisted on my having £10 quarterly added to my income, which I certainly should not have accepted if I had not been in a panic for my friends at Hanover, which had just then been taken by the French.

* * * * *

FROM MISS HERSCHEL TO J. F. W. HERSCHEL.

HANOVER, *Nov.* 1, 1824.

DEAREST NEPHEW,—

Your welcome letter, dated Slough, Oct. 22nd, had not only the most beneficial effect on my spirits, but gave the greatest pleasure to the whole family, for I find Grosekopf had been under great apprehension for your safety from the many reported accidents among the shipping on the English coasts. Count Münster, it is said, lies dangerously ill in consequence of the fright he suffered on his passage (his lady and his children were with him), and Grosekopf imagined he must have left Calais at the same time with you. But, thank God, all is well! All I meet with lament your leaving us so soon. Gauss has been here, and they say he was quite inconsolable at having missed you. Hauptmann Müller was charged with compliments, which he intends to deliver himself if I will give him leave. To be sure! and Olbers, whom Dr. Mürz saw in Bremen, was sorry not to have seen you, as you had been so near. The Duke of Cambridge, whom Dietrich met in the street, asked about you, but we could not trace you farther than Antwerp. I believe half Hanover would have been gratified if you could have made a longer stay with us. Dr. Grosekopf will one day come to England I am afraid, and talk you deaf; he is, however, a very good sort of man, and desires me to tell you that if you wanted any books you might command him, he would send you anything you wanted.

What gives me the most pleasure in reading over your letter, is your telling me that your dear mother is not in the least altered in her looks, and that she has been so considerate as to give me in her own handwriting the assurance that you are *extremely* well. That I may yet often hear the same, wishes your most affectionate aunt,

CAR. HERSCHEL.

P.S.—[To Lady Herschel] . . . .
My knowing so well to what noble purposes an experimental philosopher may use his fortune, it would make me very unhappy if my dear nephew was cramped in his. And if I could do any good by relinquishing my annuity I would leave Hanover and live on my pension in the country most willingly, and am only sorry that I have no other means of showing the care and affection I have for my dear nephew. But I beg no other notice may be taken of all I have written than often—when my nephew or yourself cannot write—to inform me by the hand of Miss B—— of all your joys and sorrows, that I may, though at this distance, sympathise with the same.

If my nephew cannot be easily supplied with the Berliner Jahrbuch, I beg he will let me know, for I have got them by me, and can send them by the messenger in January.

*　　*　　*　　*　　*

FROM J. F. W. HERSCHEL TO MISS HERSCHEL.

LONDON, *December*, 1824.

DEAR AUNT,—

My mother and self received your welcome letter, and so far from finding, as you seem to fear, the details you enter into tedious, I assure you we found them highly interesting. The sacrifices you have individually made for your family are above all praise. It would ill become me, who am a rich man (I mean in that sense only in which any man can truly be called rich,—having enough to satisfy all my moderate and rational wants), to deprive you of any, the smallest part of your income. On the contrary, it would rather be my duty, were it insufficient, to add to it, but the account you give of your situation, corroborated as it is by what I have myself seen of it, sets at rest all apprehensions on that score.

*　　　*　　　*　　　*　　　*

I hope the Catalogue of Nebulæ goes on as you wish. I shall have little time now for astronomical observations, being become a resident in London in consequence of taking on myself the duties of Secretary to the Royal Society.

*　　　*　　　*　　　*　　　*

I have sent the lenses you wished for, and also two prints of the king and queen of the Sandwich Islands, which I would be much obliged to you if you would transmit to Prof. Blumenbach, with my compliments. They are the best that have appeared, and are considered striking liknesses.

MISS HERSCHEL TO J. F. W. HERSCHEL.

HANOVER, *Jan.* 14, 1825.

MY DEAREST NEPHEW,—

*　　　*　　　*　　　*　　　*

I am now writing out the Catalogue of Nebulæ, and am at zone 30°, and hope to finish it for the Easter messenger; but my health is so wretched that I often am obliged to lay by for a day or two. Dr. Grosekopf desires his compliments, and I am to tell you that when next you come to Hanover again he can not only procure you a sight of Leibnitz's MS., but leave to take some home with you. I am in quest of a good print of Leibnitz for you, and hope soon to hear of one, which shall accompany Dr. Franklin's, which Dietrich lately found among his music.

Graf Rapfstein brought me lately the *Moniteur* of December, containing the history of your dear father's life, as read in June, etc., at full length. It is the only copy of the Court paper coming here at Hanover to the French Ambassador, and I was obliged to return it to the same; but Grosekopf has promised to procure these copies from Paris, that we may all have one. Miss Beckedorf read it to me by way of translation, and we both cried over it, and could not withhold a tear of gratitude to the author for having so feelingly adhered to truth in the details of your dear father's discoveries, etc. . . . . .

But if I have understood Miss B.'s translation right I could point out three instances where too great a stress is laid on the assistance of others, which withdraws the attention too much from the difficulties your father had to surmount.

(1.) The favours of monarchs ought to have been mentioned, but once would have been enough.

(2 & 3.) Of Alexander and me can only be said that we were but tools, and did as well as we could; but your father was obliged first to turn us into those tools with which we could work for him; but if too much is said in one place let it pass; I have, perhaps, deserved it in another by perseverance and exertions beyond female strength! Well done!

With compliments to all friends, particularly Mr. and Mrs. Babbage,

I remain, my dearest nephew,

Yours most affectionately,

Car. Herschel.

Poor Sir William Watson! [whose death had lately been announced to her.]

MISS HERSCHEL TO J. F. W. HERSCHEL.

HANOVER, *March* 7, 1825.

The birthday of my dear nephew! who I wish may enjoy in health and prosperity many returns of this day. I will drink your health, and on the 16th of this month you may return the compliment, for then I shall have completed my seventy-fifth year.

I received the parcel, not till the last day of February, which contained your letter of December 4th, with the prints of the King and Queen, which I delivered to the Regierungs-rath B——, to forward to his father at Göttingen.

The first part of your letter is filled with expressions of the most feeling kindness towards me, and I will pass them over without attempting to describe what I felt on reading the same, and merely for yours and your dear mother's satisfaction I will answer as in the way of business all you wished to know. November 22nd I received the £50 Lady H. paid over for me to Mr. Goltermann, for which I returned the day after (23rd) the formal receipt in a letter to your mother, and hope it may not have been lost (for I generally write what comes uppermost) . . . . I am ready with the Catalogue of Nebulæ, and have only to write, *not a Preface,* for I shall write what I have to say at the end . . . . I wish, in case you were not on the spot to receive the box from Mr. Goltermann yourself, you would before you left town beg Mr. G. to keep it *till you called for it yourself;* for I must confess that from the day I let the *eight manuscript books and catalogue of Nebulæ, and cata-logue of stars* drawn out of the eight books of sweeps, go out of my hands, I shall have no peace till I know they are safe in *your own,* where they ought to be. If you can think of anything else I can send you, I beg you will let me know, for a large parcel is no more trouble than a lesser one to put up. But I shall write again when I have packed up the box, and if you still wish for relics of your dear father's hand-writing, I have a great mind to part with his pocket-book (*to you only*), which he used before we left Bath. There are only a few pencil memoranda, but they show that music did not *only* occupy his thoughts, but that timber for the erection of the thirty-foot telescope of which the casting of the mirror was pretty far advanced was thought of.

But now I must say a few words to your dear mother, but I wish soon to hear that you have received this, and also a letter I sent from here on the 14th January. I hope it is not lost.

I am not very well pleased with my English, but have no time to write what I have to say over again, but this I hope you will be able to understand—that

I am

Ever your most affectionate aunt,

CAR. HERSCHEL.

FROM MISS HERSCHEL TO LADY HERSCHEL.

HANOVER, *March* 8, 1825.

MY DEAR LADY HERSCHEL,—

I received your letter of the 4th December, and it relieved me of much anxiety I felt from a fear that the subject of my long letter of November 8th might have injured me in your or my nephew's opinion, and I had nothing to console me in this uncertainty, but a line from Mr. Goltermann that he had seen you in good health and received £50 from you, which I received the 22nd November here at Hanover, and sent my thanks and the usual receipt the next day. But still I remained in uncertainty, till by a letter from Miss B. of 15th December, you kindly sent me your thanks for the very letters which caused me such fears.

But it grieves me you should yourself take the trouble of writing to me ; the least kind expression from you dictated to Miss B. is sufficient to make me happy for many days after. I hope she will not be taken from you again for a long time, for she is the most cheerful companion in health and consoling one in sickness you could have about you.

I was sorry to hear by a letter from Mr. H. Griesbach to my brother that you had had another attack of the gout, but God grant ·I may hear soon it may have been of short duration. Daily we come to hear of the departure of a friend or some one we know, but at our time of life it cannot be otherwise, for many of those we knew were older than ourselves, and it is painful to see when we at last are left to stand (or lie) alone, which is often the case with a single person ; for no attention can equal or be more cheering than what comes from the heart of an affectionate child. But no more of this ; if we must grieve, there is the comfort we shall not grieve much longer.

The death of my eldest nephew I lament sincerely, for he was deserving to have enjoyed the prosperity of his children some years longer, but by a letter I had from Miss G. I was gratified to know that they had found (for the present) so noble a support from the King and from the excellent Countess of Harcourt. As to the exit of poor F. Griesbach, it gave me more joy than pain ; for nothing but the grave could relieve him from wretchedness ; and nothing but that would rouse his posterity to a sense of their duty, which is to work for an honest livelihood ; even the youngest is old enough to do so, and I hope to hear that they may awake from their dreams of commissions in the army and midshipmen in the navy. The lot of the children of a poor musician and descendants of a menial servant (even to a king) is not to look too high, but trust to his own good behaviour and serving faithfully those who can employ them; then they will not want encouragement.

This is the way I compose myself, for help I cannot anybody any longer, and it hurts me, for I am too feeble to think much of these kind of things. The 4th April goes the messenger, and my nephew will receive my handy works and a few little publications. I have yet some publications to make which will take me some time, to go with the catalogue : and then I shall have nothing to put me in mind of the hours I spent with my dear brother at the telescopes, and for that reason I keep the five printed vols. of my brother's papers, and read them over once more before I send them to my nephew, and besides, it would be too much at once, for books are heavy.

Farewell, my dear Lady H., and remember me to Miss B., who, I hope, will be good to me and write often to

Your affectionate sister,

CAR. HERSCHEL.

P.S.—Mr. H—— is released from his plague, for his wife is dead.

MISS HERSCHEL TO J. F. W. HERSCHEL.

HANOVER, *March* 27, 1825.

MY DEAR NEPHEW,—

I hope the MS. Catalogue of Nebulæ and that of the stars, which have been observed in the series of sweeps along with the eight volumes from which they have been drawn out, will not unfrequently be of use to you.

The gauges were brought immediately after observations into a book called "Register of Star Gauges," which was kept with the "Register of Sweeps." Observations and remarks on various subjects will often be found as memorandums, made during or at the end of a sweep, to which the general index may serve as a direction—as for instance under the head of zodiacal lights—the index points out twelve different sweeps in which they were observed.

N.B.—Let it be remembered that the memorandums in the transcript of the sweeps between ‖——‖ are mine, and must be confided in accordingly.

At the end of the Catalogue of Nebulæ I have put a list of memorandums to the catalogue of omitted stars, and index to Flamsteed's Observations, contained in his second vol. They are properly not all to be called errata, but mem. of errors, which could only be solved by later observations, &c., &c.

*   *   *   *   *

All your father's papers from the Phil. Trans., which are bound in five volumes, and in which I have carried all corrections (in the Catalogues of Nebulæ) I could find, I must keep a little longer, but they shall come safe to your hands—along with Bode's and Wollaston's catalogues, when my eyes have robbed me of the pleasure of reading—for which misfortune I am in daily fear.

I am, dear nephew,

Yours affectionately,

C. HERSCHEL.

FROM J. F. W. HERSCHEL TO MISS HERSCHEL.

April 18, 1825.

DEAR AUNT,—

I received this afternoon your most valuable packet containing your labours of the last year, which I shall prize, and more than prize—shall use myself, and make useful to others. A week ago I had the twenty-foot directed on the nebulæ in Virgo, and determined afresh the right ascensions and polar distances of thirty-six of them. These curious objects (having now nearly finished the double stars) I shall now take into my especial charge—nobody else can see them. I hope very soon (in a fortnight or three weeks) to be able to transmit to you and to MM. Gauss and Harding our work (Mr. South's and my own) on the double stars, in which you will find some of my father's most interesting discoveries placed beyond the reach of doubt. It will contain measures of the position and distance of 380 double stars. But Mr. South, who is an industrious astronomer (almost as much so as yourself), has just sent me complete and accurate measures of 279 more, making in all 659. Among these we have now verified not less than seventeen connected in binary systems in the way pointed out by my father, and twenty-eight at least in which no doubt of a material change having taken place can exist. M. Struve, at Dorpat, and M. Amici, in Italy, have also taken up the subject of double stars, and are prosecuting it with vigour.

I am particularly obliged to you for my father's letters and pocket-book—they are to me a real treasure. The style of the *Éloge* in the *Moniteur* is very inferior to what I expected from Fourier; but on the whole it contains nothing materially untrue. The publications enclosed were very acceptable. I wish my uncle had not confined himself to a mere catalogue of insects, but had told us a little of their habits. Of Leibnitz's MSS. more hereafter. . . . .

The *mettwursts* * are excellent. The packets to my mother and Mary shall be sent. . . . .

Your affectionate nephew,
J. F. W. HERSCHEL.

MISS HERSCHEL TO J. F. W. HERSCHEL.

HANOVER, *May* 3, 1825.

MY DEAR NEPHEW,—

I must content myself with only writing a few lines by way of thanking you for your very interesting letter, which has taken all the care from my mind which I felt for the fate of the MS.

Before the box left Hanover, I received a very kind letter from Hofrath Blumenbach, in which was one enclosed to you; I hope it is come to hand, though I am still in doubt about your direction, and for that reason kept the letter near a fortnight before I parted with it.

You give me hope of receiving some of your and Mr. South's works for Gauss and Harding. I know no way of sending them than through Mr. Goltermann by the quarterly messenger, and that it will be well for you to make some inquiry beforehand about the time he is likely to leave England.

The Duke of Cambridge will, within a month, be in England; perhaps you will meet with him; he is a great admirer of you. Last Saturday, between the acts of the concert, he asked me many questions about you. I wish I had had your letter two days sooner, I should then have known better how to answer him. He enquired if you were much engaged with astronomy? I said you were a deep

---

* Mettwurst is a meat sausage for which Hanover is famous.

mathematician, which embraced all, &c., ... then he asked if you studied chemistry? answer, very much! you had built yourself a laboratorium at Slough, had a house in town for three years, was secretary of the Royal Society, would probably, in the vacation, be at Slough, &c., &c., and in return he told me that he heard from everybody you were a very learned philosopher; and if I tell you that the Duke of Cambridge is the favourite of all who know him, I think I have made you acquainted with one another.

My brother intends soon to write a few words about insects himself, which is almost the only object with which he *amuses* himself. It is well he does not see the word *amuses*, for I suppose it should be *sublime study*, for whenever he catches a fly with a leg more than usual, he says it is as good as catching a comet! Do you think so?

Perhaps I may have soon an opportunity of sending by Mr. Quintain a German translation of Baron Fourier's "Forlesung." I must examine first if I have the whole or not; it does not seem bad, but as I do not understand French, which I had only read to me by Miss Beckedorff, I can be no judge; but I think you will not be displeased with it; but at the ending they have not mended it, for it also says *I* had published all your father's papers, though nobody will or does believe that; still I would rather that nothing at all had been said about me than say the thing which is impossible; and I shall only fare like Bruce when he pretended to have made the drawings to his publications himself; his having wrote the book, or even having been in Abyssinia, was disbelieved.

I must only add that I am, my dearest nephew,
                    Your affectionate aunt,
                                CAR. HERSCHEL.

MISS HERSCHEL TO HERR HOFRATH UND RITTER GAUSS.

HANOVER, *Sept.* 8, 1825.

SIR,

I am almost at a loss how to express my thanks sufficiently for the kind visit with which you honoured me when last in Hanover, for not only the wish of seeing the man of whom I so often had heard my late brother speak in the highest terms of admiration has been at last gratified, but I flatter myself of having found in you, sir, a friend who will do me the kindness of presenting the works of Flamsteed (published in 1725, with my Index to the Observations contained in his second volume) to the Royal Observatory of the Royal Academy of Göttingen.

The regret I feel at the separation from books which have afforded me so many days interesting employment will be greatly softened by knowing that, referring to the memorandums in the margin of the pages in Flamsteed's second volume, much time may yet be saved to any astronomer who wishes to consult former observations, and therefore I hope you will pardon the trouble I am thus giving you, and, with the greatest esteem, believe me,

Sir,

Your most obliged and humble servant,

CAROLINE HERSCHEL.

MISS HERSCHEL TO J. F. W. HERSCHEL.

HANOVER, *Sept.* 20, 1825.

* * * * *

. . . . I know not how it comes that I am so barren of subjects for filling up these pages; my spirits are rather depressed at present on account of my brother's health, who suffers very frequently much from weakness, so that to combat against infirmities and peevishness (the usual com-

panions of old age) depends entirely on my exertion to bear
my share without communication, for unfortunately we are
never in the same mind, and with a nervous person of an
irritable temper one can only talk of the weather or the
flavour of a dish, for which I care not a pin about. But I
think I shall do well enough, for I am a subscriber to the
plays for two evenings per week, and Thursdays and Satur-
days two ladies with long titles are *at home*. This is what
they *imagine* (I believe) a learned society, or blue-stocking
club, of which, to make it complete (for all what I can say),
I must make one. I am to have a day too, viz., Tuesday,
and I begin to tremble for the end of October, when we are
to start, for in the morning I cannot work, and if I gad
about all the evenings nothing will be done. But we shall
see! one thing I must not forget, there are no gentlemen of
the party to set us right; but luckily not much is required,
—to talk of Walter Scott, Byron, &c., will go a long way;
and I subscribe to an English library, where they have all
the monthly reviews and Edinburgh Quarterly, Scott's
works, and a few other novels. . . . .

Believe me yours affectionately,

C. HERSCHEL.

FROM J. F. W. HERSCHEL TO MISS HERSCHEL.

SLOUGH [*after July*], 1825.

DEAR AUNT,—

I have sent by Mr. Golterman several volumes of
Mr. South's and my paper on double stars, which form the
third part of the Philosophical Transactions for 1824. You
will, I have no doubt, be gratified to hear that the French
Academy of Sciences have thought so well of this work as to
give us the prize of astronomy for the present year (a large
and handsome gold medal to each of us). Our competitors,
it is whispered, were Bessel, Struve, and Pons, the first for

his immense catalogue of stars; the second for his observations, also of double stars; the third for his discovery of twenty or thirty comets. Will you, on receiving them, distribute them as follows:—1. Keep the bound copy for yourself; 2. My uncle; 3. M. Harding; 4. M. Gauss; 5. The Royal Society of Göttingen. The three last, I have no doubt, M. Blumenbach will forward. I was gratified some time back by a short note from Professor Blumenbach, from which I find he received the pictures safely.

*   *   *   *   *

I have already found your Catalogue of Nebulæ in zones, very useful in my twenty-foot sweeps, and I mean to get it in order for publication by degrees; but it will take a long time, as it will require a great deal of calculation to render it available as a work of reference.

The permission to examine Leibnitz's MSS. will be very acceptable to me should I again visit Hanover, but of that I have no immediate prospect. A very intimate friend of mine, Mr. James Grahame, talks of taking up his residence at Göttingen for the sake of the library of the University. He is writing a history of America. I shall give him a letter to Professor Blumenbach, and shall beg you to introduce him to his son, Regierungsrath B., and perhaps Dr. Groskopff will make him acquainted with Dr. Koch, of the Royal Library at Hanover, who may be able to assist him in his researches. . . . . If there is anything in England you wish for, or that you cannot get so well in Hanover, pray name it, and I will make a point of procuring it. . . . .

J. F. W. HERSCHEL TO MISS HERSCHEL.

Devonshire Street, *Dec.* 30, 1825.

*   *   *   *   *

I have not been doing much in the astronomical way of late—but, *en revanche*, Mr. South has been hard at work, and has sent a second paper of 460 double stars to the Royal Society. He is returned from Paris, and is now busy erecting an observatory, as he means to stay six months in England, and cannot be so long without star-gazing. I enclose a little thing which I published in Schumacher's *Astronomische Nachrichten* which may interest you. Shortly I shall have the pleasure to transmit you some papers on the longitude of Paris, and on the parallax of the fixed stars, which I have now in hand. Do not suppose that I pretend to have discovered parallax, but if it exists to a sensible amount, I think it cannot long remain undiscovered if anybody can be found to put into execution the method I am about to propose, and I hope it will be taken up by astronomers in general.

I have so far perfected the system of sweeping with the twenty-foot that I can now make sure of the polar distances of objects to within 1', and their right ascensions to certainly within 2" of time. I have re-observed a great many of the nebulæ, and in the course of the few sweeps I have made, have discovered many not in your most useful catalogue. But I am now fixed in town for the winter, and have brought up the said catalogue to consider of the best mode of preparing it for publication, if it meets with your approbation.

Mr. South's later observations strikingly confirm the results obtained by us jointly respecting the revolving stars, and afford new and very remarkable instances in support of my father's ideas on this subject. Of one pair (the double star ξ Ursa Majoris) I have no doubt we shall soon obtain elliptic elements.

The following is the answer from Professor Gauss to the letter already given :—

DEAR MADAM,—

Being returned hither a few days ago from a journey that had kept me absent during a month, I found your favour of September 8th, together with your extremely valuable present of Flamsteed's "Hist. Cœl.," "Atlas Cœl.," and your own catalogue. Be assured that I acknowledge your kindness with the most sincere gratitude, and that these works, so precious by themselves, but much more so by the numerous enrichments from your own hand, shall always be considered as the greatest ornament of the library of our Observatory.

I am very sorry that my absence from Göttingen has deprived me of the pleasure of seeing Mr. Grahame, who was calling upon me the same day I had set out for my journey. However, I am glad to understand from your nephew's letter, which Mr. Grahame has left here, that this gentleman intends to return to Göttingen in the next year.

I cannot express how much I feel happy of having made the personal acquaintance [of one] whose rare zeal and distinguished talents for science are paralleled by the amiability of her character, and I flatter myself that in future, if I find once more an opportunity of staying in Hanover, I shall not be denied the permission to repeat personally the assurance of the high esteem with which I am,

Dear Madam,

Your most obliged humble servant,

CHARLES FREDERICK GAUSS.

GÖTTINGEN, *Sept.* 28, 1825.

CHAPTER VI.

LIFE IN HANOVER—*continued.*

MISS HERSCHEL TO J. F. W. HERSCHEL.

Feb. 1, 1826.

MY DEAREST NEPHEW,—

On the 17th January I received by the same post your letters of December 30th and January 9th. I should have answered your precious communication of December 30th immediately if I was not in hopes of receiving daily an answer to what I sent on the 28th December. I cannot express my thanks sufficiently to you for thinking me worthy of forming any judgment of your astronomical proceedings, and am only sorry that I cannot recall the health, eyesight, and *vigor* I was blessed with twenty or thirty years ago; for nothing else is wanting (and that is all) for my coming by the first steamboat to offer you the same assistance (when sweeping) as, by your father's instructions, I had been enabled to afford him. For an observer at your twenty-foot when sweeping wants nothing but a being that *can* and *will* execute his commands with the quickness of lightning [!], for you will have seen that in many sweeps six or twice six, &c., objects have been secured and described within the space of one minute of time.

I cannot think that any catalogue but the MS. one in zones (which was only intended for your own use) would facilitate the reviewing of the Nebulæ, and *you* are the only one to whom 1885, viz., 2nd and 3rd class, out of the 2500, can be visible in your twenty-foot. Wollaston, who knew this, has given in his Catalogue only 1st and 4th, &c. classes of the first 1000, the second not having been published at that time, and they are without the yearly variation.

Bode has given the first and second Catalogues complete, and calculated the yearly variation to each by de Lambre's Tables. (See Bode's preface, p. iv., line 18.) The last 500 were not published yet in 1800, or rather 1801. I only mention this that if you wanted the variations, and had a mind to trust to that catalogue of errors, it would save an immense trouble by copying them. But the more I think of these, the more I doubt if it would not be in-juring the places of objects merely (though accurately) pointed out, to calculate them in the same manner as stars repeatedly observed in fixed instruments ; and I doubt if your father noticed Bode's having done so.

You will find undoubtedly many more nebulæ which may have been overlooked for want of time, flying clouds, hazi-ness, &c., especially in those sweeps which are registered *half sweep.* It is a pity time could not be found for making, as was often intended, a register in which the boundaries of the sweeps, with the nebulæ, were all brought to one time, either to Flamsteed's or 1790 or 1800. The register in Flamsteed's time, which is from 45° to 129°, is for that reason the best mem. At the time that register was made, the apparatus for sweeping in the zenith was not completed, and higher than 45° was not used.

If you should wish in the latter part of the summers (when your father was generally from home) to fill up the unswept part of the heavens, you might perhaps discover as many objects as would produce a pretty numerous catalogue. You will see in the register of Flamsteed's time a curved line which denotes that the Milky Way is in those places, and if you see an *L* and find a cluster of stars thereabout, I shall claim it as one of those I mentioned in my last letter to you. It was the assistant's business to give notice when such marks or any nebulæ in the lapping over of the sweep either above or below were within reach, by making the workman go a few turns higher or lower. (N.B. No more than is convenient without deranging the present sweep.) But I am forgetting myself, and fear I am tiring you unnecessarily, and will only add that if your father wanted at any time to review or to show any of his planetary or other remarkable nebulæ to his friends, the time and P.D. was, by the variation of its nearest star in Wollaston's or Bode's catalogue, brought to the intended time of observation, and P.D. — comp. of latitude, with allowance for refraction, gave the quadrant for setting the telescope.

But after all, dear Nephew, I beg you will consider your health. Encroach not too much on the hours which should be given to sleep. I know how wretched and feverish one feels after two or three nights waking, and I fear you have been too eager at your twenty-foot, and your telling me that you have been unwell for some months, and now only begin to feel better, makes me very unhappy, and I shall not be comfortable till I see by your next that you are perfectly well again; I am quite impatient to see what you have to say about the parallax of the fixed stars, but on such occasions I am vexed that your father did not live to know of your grand discoveries. You say something of a paper on the longitude of Paris; I hope you will think of Gauss when you have anything new.

Among the letters from your father's correspondents in alphabetical parcels you will find under the letter P. some of Pond's, who was about the end of the last century in Lisbon, with an excellent seven-foot telescope of your father's, and I remember that several letters passed between them about a double star in Böotes.

I am much obliged to you for the sheet of Schumacher's " Astronom. Nachrichten." It is highly interesting to me, and will set many a one right without offending anyone. On looking in the 2nd Catalogue of double stars, No. 104, ζ Böotes, VI. Class, November 29th, 1782, and 3rd Catalogue, No. 114, ζ Böotes, I. Class, April 5th, 1796, I cannot help thinking on the possibility that in the lapse of thirteen years and a half the small stars may have come out from behind the large one. But I beg do not laugh at me for breaking my head about these things, and I will now begin to talk about what I can comprehend.

From your mentioning Mr. South in your last letter, I fear he intends leaving England, at which I should be very sorry on your account, for if I should not live long enough to know you comfortably married, I could only console myself by your having always a Babbage, South, or Grahame to pass your social hours with. If you can meet with a *good-natured, handsome, and sensible young lady,* pray think of it, and do not wait till you are old and cross. And let me know in time that I may set hands to work to make the bridal robe ; here are women who work exquisitely, and at a price within the reach of my purse.

    *   *   *   *   *

P.S.—Dear Nephew, I have spent too much time in gossiping with your dear mother for saying anything besides, but I am,

Your most sincere and affectionate aunt,

CAR. HERSCHEL.

MISS HERSCHEL TO J. F. W. HERSCHEL.

HANOVER, *Aug.* 8, 1826.

*   *   *   *   *

The long continuance of the great heat has had so very bad an effect on my feeble frame; and considering my advanced age, I ought not to put off the making a sort of a will, which I would set about with the greatest pleasure if I had anything to leave for which you would be the better. But I am sure you will not be disappointed, for you remember I parted with my little property before I left England (against your good advice) because I thought at that time I should not live a twelvemonth.

*   *   *   *   *

From the first moment I set foot on German ground, I found I was alone. But I could not think of separating myself from him, [her brother Dietrich] especially as his health is so very precarious, that I often think he will go before me. At this present moment he is in bed very ill, suffering from weak nerves. But the above is all by way of showing you the necessity for begging you to answer to the following questions.

My sweeper I wish to leave to Miss Beckedorff, and the picture of the Princess of Gloucester to her mother, for the two ladies have been my guardian angels for many years.

Dr. Groskopff is to have the seven-foot reflector, though I know it will only be a relic to him, but it will not be destroyed or sold for an old song. My clothing and such articles of furniture as I have been obliged to purchase, my three nieces may divide themselves in. Your dear father's publications in five volumes, Bode's and Wollaston's Catalogues (full of my memorandums), and one of my Indexes, shall be sent to you. Also a rough copy of the general Index to your father's observations, and several articles of that sort with memorandums taken from what I have called a Day-book, which at leisure you may look over and afterwards consign to the flames, for I cannot take it in my heart to do it myself.

The observations on double stars by you and Mr. South (so handsomely bound) and the volume sent last, by South, shall I send them to you?—else I leave them to the Duke of Cambridge!—answer required.

Taylor's tables, will they be of use to you for your godson Babbage?—else they must be only an ornament to Groskopff's library!—answer required.

I am impatient to have your answer to this stuff, which I am almost ashamed to trouble you with.

*   *   *   *   *

My next shall be of a more agreeable subject, and I have only to say,

I am,
Your most affectionate aunt,
C. HERSCHEL.

FROM J. F. W. HERSCHEL TO MISS HERSCHEL.

MONTPELLIER, *Sept.* 17, 1826.

DEAR AUNT,—

You will think me a strange gad-about, but my last, if you have got it, will have prepared you to expect a letter from either the north or south of Europe from me, in short from any country except England. I was then not decided whether to go to Norway or the south of France, but here I am at last, and having a letter-writing day before me and yours of the 8th August in my portfolio, I cannot do better than to answer it.

With regard to the dispositions you mention in your letter, and respecting which you express a wish for my

opinion, they are such as it is impossible to do otherwise
than approve, and such as the good sense and kindness
which marks everything you do has dictated.

*   *   *   *   *

I have been rambling over the volcanoes of Auvergne,
and propose before I quit this, to visit an extinct crater
which has given off two streams of lava at Agde, a town
about thirty miles south of this place on the road to the
Spanish frontier.   Into Spain, however, I do not mean to
go,—having no wish to have my throat cut.   I am told,
however, that a regular diligence runs between this and
Madrid, and is as regularly stopped and robbed on the
way.

You say you wish for an answer respecting the vol. of
observations on double stars, sent by Mr. South and myself, but can I do better than leave such matters to your
judgment ?   At the same time, as having belonged to you
they could not but have a value in my eyes beyond my own
copy ; but pray decide yourself.   I have several left.

I regret extremely to hear you feel those little (perhaps
not little) inconveniences we are none of us exempt from,
arising from the imperfections of human nature, both in
ourselves and those we live with.   I believe the best receipt
for them is endurance and a determination to show ourselves
superior to them.

I have my rubs now and then too, but I make up my mind
to them as quite inevitable, and arising from causes over
which I have no control.   I am very sorry to hear of my
uncle's bad state of health.

I must be in England in the beginning of October, or at
farthest by the 15th.   So, you see, I have no time for
Hanover on my way back.   It is dreadfully hot here, and I
am much disappointed with the place.   However, I hope to
get one day of intense sunshine while I remain in this

latitude on account of some observations on solar radiation
I have to make with a new instrument which I made before
I left England, and brought with me. I carried it up the
Puy de Dome, and was in hopes to have used it at the Great
St. Bernard, in Switzerland, but that must now stand over
for another year.

Adieu, dear aunt, and believe me—" where'er I go, whatever realms I see "—

Your affectionate nephew,
J. F. W. H.

MISS HERSCHEL TO J. F. W. HERSCHEL.

Sept. 29, 1826.

My dearest Nephew,—

Within this hour only I received your dear letter,
dated Montpellier, Sept. 17th, which I assure you has made
quite another (and what is more) a proud woman of me ; for
your answers to my few questions are so kindly expressive of
approbation, that I shall in future not fear to follow my
own opinion, which through my whole lifetime I never ventured to do before.

I am glad you did not come to Hanover, for I am sure to
part from you once more would finish me before I am quite
prepared for going.

The letter you mention having written me before you
left England I have not received. The fault does not lie
here, for the secretary here takes too much pleasure in
sending me my letters.

I must hasten to get my packet away, but will only beg
to let me know through Miss Baldwin as soon as you get
home of your safe arrival, for I fear you must often be
exposed to great dangers by creeping about in holes and

corners among craters of volcanoes, but you know best, and I hope you found something. . . . .

I am,
My dear nephew's affectionate aunt,
C. Herschel.

MISS HERSCHEL TO J. F. W. HERSCHEL.

Nov. 1, 1826.

My dearest Nephew,—

The 1st vol. of the translation of your dear father's papers is come out. I shall have it in a few days from the bookbinder, and in February, I am told, the next volume will make its appearance. I wish you would inform me *as soon as possible* if I shall send you a copy, that I may write for one in time to have it ready by the end of December, when the messenger leaves Hanover. It is a pity you cannot have it immediately. The plates are not with the work, but are to be had bound in a separate book (I suppose when the whole is finished).

I long to know that you are arrived safe and in good health in England again, for by your last, dated Montpellier, Sept. 17th, I see that you had then another volcanic mountain to visit, besides an observation to make on solar radiation with your new instrument; the very thought of it puts me in a fever all over—at this present moment, though we have no longer to complain of heat; so I beg you will inform me that your health has not been injured, and that you have not been totally disappointed in your researches.

I lead a very idle life, my sole employment consists in keeping myself in good humour and not be disagreeable to others.

Groskopff tells me the translation of your father's papers

causes a great sensation among the learned here in Hanover.

       *   *   *   *   *

           Believe me, dearest nephew,
               Yours, most affectionately,
                        C. HERSCHEL.

     MISS HERSCHEL TO J. F. W. HERSCHEL.

                                 Dec. 5, 1826.

MY DEAREST NEPHEW,—

    I received your letter of the 18th November, the day before yesterday, therefore fifteen days old, which is pretty well considering the time of year. I hope this will reach you soon, for I have longed very much to give you an account of the last parcel of papers you sent, which I only deferred till I had received an account of your safe arrival in England by your own hands.

    The parcel which you gave to Mr. Goltermann on the 18th August arrived here by the messenger on the 3rd November, and five days after (which it took me to dry the copies, for the messenger had met with storm and accidents at sea, and some of his boxes had been under water), viz., the 18th Nov., I sent to Göttingen, according to direction, with a note, to Gauss. And those to Bessel and Encke I enclosed with Bode's copy, and wrote a letter to the same by way of thanks for some kind enquiries he had made after me ; and now I see that fourteen days after this good man [Bode] departed this world in his eightieth year, but I have no doubt he has delivered the papers immediately, for he had no illness, and was at his last hour at his writing-table employed with writing the " Berliner Jahrbuch " for 1830.

    The copies were, after being dried, perfectly clean, no stain remaining, and that they were so long detained is not the fault of Mr. G., for the Michaelmas messenger was the first that went after the 18th Aug. In the parcel I found also the letter you wrote before leaving England, which I concluded to have been lost, but now all is safe.

Sun and Comet.—At Hanover totally cloudy, and by what I can learn from a certain astronomical gossip, Prof. Wild, it has been so throughout all Germany, for *he* has had no account that anything has been seen on the 18th Nov. On the 17th it is mentioned (in the *Zeitungen,* I believe) a large spot on the sun to have been observed at Frankfort, but the 18th being cloudy it could not be pursued.

In your observations with the twenty-foot you mention a Mr. Ramage as having observed with vou ; and in another place you speak of his twenty-five-foot reflector. Pray tell me something about this gentleman, for I never heard his name before, and if I had not been so fortunate as to have seen Babbage and South just before I left England, I should not now have the comfort to know you had so estimable friends to communicate with ; and I shall rejoice to know that the number of valuable men I have known, and are no more, might be replaced by some who are worthy to be contemporary with the son of your father !

You ask, as it were, if I were satisfied with the way in which you have mentioned me in that paper ? If I should answer honestly I should say not quite, for you set too great a value on what I have done, and by saying too much is saying too little of my brother, for *he* did all. I was a mere tool which *he* had the trouble of sharpening and to adapt for the purpose he wanted it, for lack of a better. A little praise is very comfortable, and I feel confident of having deserved it for my patience and perseverance, but none for great abilities or knowledge. But of this you will perhaps be a judge, as I am now gathering from loose memorandums a little history of my life during the years from 1772 to 1788. . . . .

*   *   *   *   *

You mention a monkey-clock, or jack, in your paper. I would only notice (if you mean the jack in the painted deal case) that Alex made it merely to take with me on the roof when I was sweeping for comets, that I might count seconds by it going softly downstairs till I was within hearing of the beat of the timepiece on the first floor (at that time our observatory) all doors being open. Your father never used it except when polishing the forty-foot. . . . .

In about three weeks the messenger leaves Hanover, and I will send you the first volume of the translation of your father's papers; but I shall not order ten copies as you desired, till you give me further orders, for I do not think you will be pleased with the work, and it seems there is not much call for them. Dr. Luthmer, says Pfaff, was not the man who ought to have attempted such a work, it ought to have been a Bessel.

To your dear mother and Miss B. I beg to be kindly remembered,

And remain

Yours, most affectionately,

C. Herschel.

FROM MISS HERSCHEL TO J. F. W. HERSCHEL.

Dec. 24, 1826.

Dearest Nephew,—

You will with this receive the only volume of the translation (printed on *bad* paper, *without* the prints, &c., &c.,) which is out at present, and unless you desire me in your next to send you ten copies, I shall only take one which can serve us both.

I certainly will do as you desire, and tell you the amount, if at any time you should want some expensive publication, as our bookseller here can get by return of post from Leipsic whatever is ordered. But as to trifles, I beg you will never think about, as I should be at a loss for proving that you, my dearest nephew, are daily in my mind, when I am lavishing *sums* on nieces and grand-nephews, and nieces who care not for me, nor I for them. But enough of this; only write me sometimes what you and your astronomical friends are doing.

I was much gratified to hear that Mr. South had received the medal. Groskopff has seen it announced in the papers, where your name was also honourably mentioned; these are the morsels for me to feed upon, for here are no astronomers but one, Dr. Luthmer, who observes Jupiter's satellites, as you may see by the Berliner Jahrbuch, which I suppose you have, as usual, else I have got them from '23 to '29, and could send them.

I must write a line yet to your dear mother and Miss B., and will conclude with wishing you a merry Christmas and a happy New-year (as the saying is), and with loves and compliments wherever they are due, &c., &c.,

C. Herschel.

P.S.—My brother is at present tolerably well, but I hardly ever knew a man of his age labouring under more infirmities, nor bearing them with less patience than he does; the rest are well enough ! *

MISS HERSCHEL TO J. F. W. HERSCHEL.

April, 1827.

Dearest Nephew !—

I have more than once asked if you would have my history, but my question has never been answered, and I am (though unwillingly) obliged to send it off without having received your permission. . . . .

Perhaps I have told you nothing but what you have known long since; but as my thoughts are continually fixed on the

* Dietrich Herschel died towards the end of January, 1827.

past, I was, as it were, conversing with you on paper, not choosing to trust them to any one about me, for I know none who would understand me, or whom it can concern, what *my own private opinion* and remarks have always been about the transactions that continually passed before my eyes. But there can be no harm in telling my own dear nephew, that I never felt satisfied with the support your father received towards his undertakings, and far less with the ungracious manner in which it was granted. For the last sum came with a message that more must never be asked for. (Oh! how degraded I felt even for myself whenever I thought of it!) And after all it came too late, and was not sufficient; for if expenses had been out of question, there would not have been so much time and labour and expense, for twenty-four men were at times by turns day and night at work, wasted on the first mirror, which had come out too light in the casting (Alex more than once would have destroyed it secretly if I had not persuaded him against it), and without two mirrors you know such an instrument cannot be always ready for observing.

But what grieved me most was, that to the last, your poor father was struggling above his strength against difficulties which he well knew might have been removed, if it had not been attended with too much expense. The last time the mirror was obliged to be taken from the polisher on account of some obstacle, I heard him say (in his usual manner of thinking aloud on such occasions), "It is impossible to make the machine act as required without a room three times as large as this."

But when all hopes for the return of vigour and strength necessary for resuming the unfinished task was gone, all cheerfulness and spirits had also forsaken him, and his temper was changed from the sweetest almost to a pettish one; and for that reason I was obliged to refrain from

P

troubling him with any questions, though ever so necessary, for fear of irritating or fatiguing him; else there was work enough cut out for keeping me employed for several years to come, such as making correct registers of the sweeps in which all Nebulæ were to be laid down and numbered, complete Catalogues, &c. But what I most regret is, that I never could find an opportunity of consulting your father about collecting the observations made with the 40-foot into a separate book from the journals, into which they were written down among other observations made with the other instruments in the same night. I know besides that many must have been lost, being noted only either on slates or on loose papers, like those on the first discovery of the Georgian Satellites. Owing to my not being, as formerly, the last nor the first at the desk (generally retiring as soon as the mirror was covered), the memorandums were often mislaid or effaced before I had an opportunity of booking them. But I ought to remember that suchlike incomplete observations were made under unfavourable circumstances. For instance, the P. D. clock disordered by not having been used for some time; the timepiece not having been regulated, nor *every one* of the out-door motions wanting oiling or cleaning; company being present; the night not perfectly clear; and, in general, the first night the instrument is used after it has been left at rest for some time, it cannot be expected that all should go on without interruption or ease without a good mechanical workman had spent best part of the day in looking over all the motions, in doing which your father used to find great pleasure.

But what I most lament is, that between the interval before your coming to the age of forming a proper opinion of the instrument, it had nearly fallen into decay almost in all its parts. But we have all had the grief to see how every nerve of the dear man had been unstrung by over-exertion ; and that a farther attempt at leaving the work complete became impossible.

But, by the description of the forty-foot telescope given in the Philosophical Transactions, May 18, 1795, it may be seen what a noble instrument had been obtained by all the exertions described in my narrative; but from that description so briefly given there, no idea can be formed with what accuracy and nicety each part of the whole had been executed to make it an instrument fit for the most delicate observations.

P.S.—I must say a few words of apology for the good King, and ascribe the close bargains which were made between him and my brother to the *shabby, mean-spirited advisers* who were undoubtedly consulted on such occasions; but they are dead and gone, and no more of them! Sir J. Banks remained a sincere well-meaning friend to the last.

Farewell, my best Nephew!

MISS HERSCHEL TO J. F. W. HERSCHEL.

May 8, 1827.

DEAREST NEPHEW,—

Through the friendly care of Mr. D—— I am enabled to send you the first and second volumes of Locke, the third volume, I hope, will yet be found, and I shall send it by another opportunity. I know you will prize the book when you know that it was one of your father's earliest treasures, purchased out of his own little savings, at the age of 18 years*—when, along with his father and eldest brother, he was in England with the Hanoverian Guard, which you will see by the date and name, written in his own beautiful handwriting. When in 1758 he again went to England, it was under such unpleasant circumstances that he was obliged to leave it to his mother to send his trunk after him to Hamburg; and she, dear woman, knew no other wants

* See p. 10.

but good linen and clothing, and your dear father's books and self-constructed globes, &c., were left behind, and served us little ones for playthings till they were destroyed; but no more of this. You must excuse an old woman, especially such a one as your old aunt, who can only think of what is past, and is for ever forgetting the present. . . . .

*   *   *   *   *

Now, there is gone a Herr Von Münighausen, who had asked the same favour, [that of being allowed to take a parcel to England] for they are all very desirous of knowing J. H., and would have called on me, and perhaps I might have had my hand kissed once more. I assure you it is no trifle here at Hanover to have one's hand kissed, if one cannot count one's forefathers for sixteen generations back as ennobled; but, alas! he was obliged to go at a moment's warning; but Dr. Gr. gave him your address, and I hope you will receive him kindly.

Farewell, dear Nephew, &c., &c.,
C. Herschel.

J. F. W. HERSCHEL TO MISS HERSCHEL.

Between 4th and 11th May, 1827.

My dear Aunt,—

I received yesterday your packet by Mr. Goltermann, containing the ten copies of the first vol. of Pfaff's translation of my father's works—with *the plates,* which are really abominable. However, there is no help for it. I shall destroy those of the Nebulæ. A much more interesting part of its contents is your account of your own history, for which I cannot enough thank you, and it is really one of the most precious documents you could have sent me; every line of it affected me deeply. The point of view in which it places my father's character is truly noble. You under-rate both the value and the merit of your own services in his cause, but the world does you more justice, and his son feels them a great deal more than he knows how to express. I shall preserve this as the most precious thing, and you will add to the obligation you have conferred on me by sending the papers you refer to under the title of No. I.

The Journals and the *mettwursts* * also came safely ; the Journals contain some very curious matter not known in England, and which comes very opportunely here, where, I am sorry to say, science is going to sleep.

I have just completed a second Catalogue of double stars, which will be read at the Astronomical Society (of which I now have the honour to be President) on Friday (May 11th) next (if I can get it fairly copied in time). My work in the Review of Nebulæ advances slowly, as I can very seldom get a night or two at proper times of the moon and year to sweep. But I find your Catalogue most useful. I always draw out from it a regular *working list* for the night's sweep, and by that means have often been able to take as many as thirty or forty nebulæ in a sweep. I have now secured such a degree of precision in taking the places of objects in the telescope, that the settling stars (which I prepare a list of each night and arrange them in order of R. A. in the working list) cross the wire often on the very beat of the chronometer when they were expected, and not unusually enter the field of view bisected by the horizontal wire of the eye-piece. In short, I reckon my average error in R. A. in determining the place of a new object by a single observation, not to exceed one second of time, and in Polar distance a quarter of a minute. This you will easily perceive to be a considerable improvement in respect of precision, which is more my aim than it was my father's, whose object was only discovery. I have found a great many nebulæ not in your Catalogue, and which, therefore, I suppose

* Meat sausages—a Hanoverian delicacy.

are new. But I won't plague you any more with this at
present.

      *   *   *   *   *

Believe me, dear Aunt,
Your affectionate Nephew,
J. F. W. H.

MISS HERSCHEL TO LADY HERSCHEL.

HANOVER, *July* 10, 1827.

MY DEAR LADY HERSCHEL,—

      *   *   *   *   *

It makes my heart overflow with gratitude when I
see so many worthy people remember me with kindness,
and I particularly rejoice that Mrs. Morsom has borne her
misfortunes with such resignation so as to be still able to
participate in the society of her friends; of which I am,
alas! through the great distance, entirely cast out, and am
obliged to trust alone to myself for keeping up my spirits,
and to bear pain and sickness, or feel pleasure without
having anybody to participate in my feelings. Out of my
family connections, however, I can boast to possess the
esteem and love of all who are great and good in Hanover,
but to a lonely old woman, who is seldom able to go into or
receive company, this does not compensate for the want of
sympathising relations.

But I have now, by change of apartments, made myself
quite independent of anybody. As long as I can do something for myself this will do very well; but I must not
meet troubles at a distance. I may, perhaps, be spared a
long confinement before I leave this world, else such a thing
as a trusty servant is, I believe, hardly to be met with in this
city, which, along with the people in it, are so altered since
the French occupation and the return of the military with
their extravagant and dissipated notions, imbibed when in

Spain and England, with their great pensions, which they draw from the latter country, that it is quite a new world, peopled with new beings, to what I left it in 1772. Added to this comes the fear of having my new little English bed (which on my removal I made with my own hands) burnt before I am aware ; for, figure to yourself what danger one continually must be exposed to, when, in the house where I live, seven families (besides the floor my sister-in-law and I occupy) with their servants and children, are living, and their firing wood and turf is all carried over our heads. About a month before Easter a great brewery, very near us, burnt down, with many surrounding houses, to the ground. I looked out of the window, and the burning flakes fell on my forehead ; besides this, I have had four times the fright of fires at some greater distance.

*   *   *   *   *

Your most affectionate Sister,
C. HERSCHEL.

MISS HERSCHEL TO J. F. W. HERSCHEL.

Aug. 16, 1827.

MY DEAREST NEPHEW,—

On the 9th I received the papers with your short but sweet letter, and according to your direction they are by this time at their destined places, all but Struve's and Bessel's ; the latter, I was obliged to leave to the care of Encke, and Struve's to Schumacher. I am particularly obliged to you for your second Catalogue of double and treble stars, which on reading it once over, makes me long for the time when I shall be perfectly at ease to take it up again ; for, by the manner in which you gentlemen now attack the starry heavens, it seems that there will soon remain nothing to be discovered.

You mention that Mr. Baily intends to bring Flamsteed's

omitted stars into a Catalogue; I send you a few errata, as I am not sure of having carried them into the copy I left with the three volumes of Flamsteed's works. And in the list of your father's MS. papers, in the packet "Auxiliary Article," is a Catalogue of omitted stars arranged in order of R. A. (a copy of one which I gave to Dr. Maskelyne in 1789). This, may perhaps save some trouble to Mr. B. in arranging them.

Some time ago Count Kupfstein sent me a copy of Littrow's observations to look at (Part VII. of forty-three sheets large folio), which he publishes at the order and expense of the Emperor. The copy was for the University of Göttingen; but I could only admire the fine paper and beautiful print, as I do not understand the manner in which observations are made with the new invented instruments, for at the time I made a fortnight's visit to Greenwich, in 1798, they had only the mural quadrant and the meridian passage instruments.

I must conclude for want of time; and, to say the truth, I am fatigued, for I cannot sit up for any length of time, till eight or nine o'clock in the evening, when I find myself always the most fit for society, or a little business. The weather has been too warm for me, and I have done nothing but sleep in the mornings and afternoon, and the worst is that everybody goes to bed between nine and ten, and then I have no society but those I can meet with in a novel. The few, few stars that I can get at out of my window only cause me vexation, for to look for the small ones on the globe my eyes will not serve me any longer.

Tell your dear mother she must not give me the slip, for I will and cannot mourn for anyone more that I love.

I remain, &c., &c.,

C. HERSCHEL.

MISS HERSCHEL TO J. F. W. HERSCHEL.

Sept. 25, 1827.

DEAREST NEPHEW,—

Herewith you will receive what I have called No. 1, which was never intended to have met your eye as it is; but, as contrary to my expectation, my No. 2 was so cordially received by you, I had intended to send you only an abridgment of it, because it contains many things which must be very uninteresting and almost unintelligible to you on account of your being unacquainted with the (then) manners and customs of this country, besides requiring to remember that my father and mother were born and educated some hundred and twenty years back. But I must send it as it is, or destroy it immediately, for I feel I shall *now* never get well enough for making any alteration further than running my eye over it and adding a note here and there where necessary. But I wish not to leave my memorandums any longer to the chance of falling into the hands of officious would-be learned ignorance, to furnish a paragraph in some newspaper or journal.

I will, however, save you and myself the trouble of further apologising for sending you these papers, but just explain my reason for taking a copy of them with me.

When I took my leave of the contents of your father's library, it was parting from *all* with which my heart and soul had been engaged for the best part of my life, and I could not withstand the temptation of carrying away with me an index for assisting my memory when in my reveries I should imagine myself to be on the spot where I took leave of all that had been most dear to me.

What is contained in No. 1 I had intended for an ever-lasting pleasing melancholy subject for conversation with my brother Dietrich, if I should go back again to the place where I first drew my breath, and where the first twenty-two years of my life (from my eighth year on) had been sacrificed to the service of my family under the utmost self-privation without the least prospect or hope of future reward. Or in case I had died in England, it was to have been sent to D., for I wished him to get a more correct idea of our father than what I thought he had formed of that excellent being.

He never recollected the eight years' care and attention he had received from his father, but for ever murmured at having received too scanty an education, though he had the same schooling we all of us had had before him.

I ought to remember here, I suppose it was in the year 1818, or perhaps earlier, your father wished to draw up the biographical memorandum you have in your possession. But finding himself much at a loss for the dates of the month, or even the year when he first arrived in England with his brother Jacob, I offered to bring some events to his recollection by telling what I remembered having passed at home during the two years his brother was with him, with the proviso not to criticize on telling my story in my own way. But not being very positive about the exact date when my eldest brother returned, I wrote to my brother D. for the date when Jacob entered the orchestra, and found not to have been much out in my reckoning. And from that time on, your father became more settled, and could have recourse to the heads of his compositions, &c., &c., for the dates he wanted for his purpose.

Of all that follows I do not remember to have shown him a single line. But as I had once begun the subject I did not know how or where to leave off, and went on, thinking my brother D. might some time or other profit by getting better acquainted with what had passed in our family before his time, and during his infancy, till the death of his father, which happened when D. was in his twelfth year, of which, from the conversation I had with him during the four years between 1809 and 1813, when last in England, I found he had not the least notion, or had purposely formed a very erroneous one.

But in the last hope of finding in Dietrich a brother to whom I might communicate all my thoughts of past, present, and future, I saw myself disappointed the very first day of our travelling on land. For let me touch on what topic I would, he maintained the contrary, which I soon saw was done merely because he would allow no one to know anything but himself. . . . . Of course, about these papers I could never have any conversation with him nor anybody else, and I send them to you for your perusal, because I do not wish to keep them any longer, and you may put them in the fire after having read them over.

Adieu, dear Nephew, believe me ever,

Your most affectionate Aunt,

CAR. HERSCHEL.

MISS HERSCHEL TO J. F. W. HERSCHEL.

Dec. 22, 1827.

* * * * *

. . . . Of Dr. Olbers, I hear frequently through a sister and niece here at Hanover; the last was that he was lamenting at Captain Müller not having brought the paper you had intended for him ; the poor man, I hear, is grown corpulent and short-breathed, so that he cannot mount up to his observatory without difficulty.

I heard from Capt. Müller (what I had been thinking before) that poor Encke has not changed his situation for the better. I do not mean with regard to income, for I believe his salary is four or five thousand thalers per year. which is equal, or even more, than that of a Prime Minister ; but he has no instruments. Much is promised, but he gets nothing; and besides, his family is settled in Götha. It is a pity such a man should be obliged to be idle.

In my last to your dear mother I wrote nearly all I had to say about myself, except what concerns my health, of which I could not give a very good account. Lately I was obliged to consult an oculist, but I suppose he cannot help me, for he has not ordered me anything. I cannot, after having been asleep, get my eyes open again for a considerable time, this is attended with a violent headache and giddiness—but no more of this.

Once you were asking me if I wanted a few of my Indexes; if it is not too late (as you have given up the secretaryship), I would be glad of a couple. N.B.—A hundred copies were promised me as a present, and were not half of them received. The one I have by me, which is intended for you, with my corrections in it, is spoilt in the binding; and I should like to give one to the Duke of Cambridge, to put him in mind of the little old woman who has so frequently been cheered by his kind attentions.

I remain your most affectionate Aunt,

CAR. HERSCHEL.

MISS HERSCHEL TO LADY HERSCHEL.

May 9th, 1828.

MY DEAR LADY HERSCHEL,

This is to be a letter of thanks, but I cannot determine to whom I am to allot the greatest portion of my thanks, to you or Miss Baldwin, for her agreeable letter of April 15th, in which so many interesting friends and acquaintances of mine are remembered. For, believe me, my dear Lady H., it is ever with great reluctance I am yearly drawing on you for so considerable a sum, which in the end must some time or other be felt by my dear nephew; but who would have thought it, that I should last so long? but now I am losing strength daily, and I cannot expect to be long for this world. I only say this by way of putting you in mind that I received my annuity at the beginning of the first half-year, and therefore when you hear of my death all your care on my account must be at an end, for I leave a sufficient sum to defray all possible expenses attending a funeral, &c.

But there is nothing grieves me more than that, at my leaving England, I gave myself, with all I was worth, to this branch of my family, believing them (from what my brother D. and their letters told me) as many noble-hearted and perfect beings as there were individuals. But though I am disappointed, I should not like to take back my promise, which could not be done without creating ill-will, and I am too feeble to bear up against any altercation.

I see I have not left room for all the loves and compliments, but I beg you will give them to whoever is kind enough to remember,

My dear Lady Herschel,
Your most affectionate Sister,
C. HERSCHEL.

In February, 1828, Miss Herschel's services to the Science of Astronomy were recognized by the presentation to her of the Gold Medal of the Royal Astronomical Society.

FROM J. F. W. HERSCHEL TO MISS HERSCHEL.

May 5, 1828.

DEAR AUNT,—

Herewith you will receive the medal, of whose award you will have read in the printed notice I enclosed you some ten days ago. My mother also begs your acceptance of a pair of bracelets, and begs me to thank you for your kind and beautiful present of needlework (which even I

could admire), and for the *mettwursts* (which I fully com-
prehended, and part of which I still comprehend, having
regaled on one for breakfast). My mother and cousin are
quite well, and desire their best love. Slough stands where
it did, and star-gazing goes on well. I have just erected a
new instrument (Mr. South's *ci-devant* large equatorial),
and you shall hear from time to time what is doing. . . . .

Your affectionate Nephew,
J. F. W. HERSCHEL.

The presentation of the medal is the natural duty
of the president of the society, but as Mr. Herschel
held that office on this occasion, and had with characteristic modesty "resisted," as he confesses, the proposed honour, the following supplemental address was
delivered by Mr. South, the vice-president, who presented the medal to Miss Herschel through her
nephew. It is an eloquent and not unworthy tribute,
and an interesting memorial of the esteem in which
she was held by the most distinguished body of
scientific men in the kingdom.

Address to the Astronomical Society, by J. South, Esq., on
presenting the Honorary Medal to Miss C. Herschel, at
its Eighth General Meeting, February 8th, 1828.

GENTLEMEN,—

Our excellent president, in his address, has informed
you of the appropriation of two of your gold medals since
our last anniversary:—a third, however, has been decreed
by your council; and when it is known that Miss Caroline
Herschel is the individual to whom it stands adjudged, it is
not difficult to determine why the president has avoided the
slightest allusion to it.

But that your Council has not selected one from the many of its members infinitely more competent to do justice to the transcendent merits of that illustrious lady is most assuredly matter of regret. I must therefore throw myself upon your indulgence, hoping that the goodness of the cause may in some measure compensate for the inability of its advocate.

The labours of Miss Herschel are so intimately connected with, and are generally so dependent upon, those of her illustrious brother, that an investigation of the latter is absolutely necessary ere we can form the most remote idea of the extent of the former. But when it is considered that Sir W. Herschel's contributions to astronomical science occupy sixty-seven memoirs, communicated from time to time to the Royal Society, and embrace a period of forty years, it will not be expected that I should enter into their discussion. To the Philosophical Transactions I must refer you, and shall content myself with the hasty mention of some of her more immediate claims to the distinction now conferred. To deliver an eulogy (however deserved) upon *his* memory is not the purpose for which I am placed here.

His first catalogue of new nebulæ and clusters of stars, amounting in number to one thousand, was made from observations with the twenty-foot reflector in the years 1783, 1784, and 1785. A second thousand was furnished by means of the same instrument in 1785, 1786, 1787, and 1788; while the places of 500 others were discovered between 1788 and 1802. But when we have thus enumerated the results obtained in the course of *sweeps* with this instrument, and taken into consideration the extent and variety of the other observations which were at the same time in progress, a most important part yet remains untold. Who participated in his toils? Who braved with him the inclemency of the weather? Who shared his privations? A female. Who was she? His sister. Miss Herschel it was who by *night* acted as his amanuensis: she it was whose pen conveyed to paper his observations as they issued from his lips; she it was who noted the right ascensions and polar distances of the objects observed; she it was who, having passed the night near the instrument, took the rough manuscripts to her cottage at the dawn of day and produced a fair copy of the night's work on the following morning; she it was who planned the labour of each succeeding night; she it was who reduced every observation, made every calculation; she it was who arranged everything in systematic order; and she it was who helped him to obtain his imperishable name.

But her claims to our gratitude end not here; as an original observer she demands, and I am sure she has, our unfeigned thanks. Occasionally her immediate attendance during the observations could be dispensed with. Did she pass the night in repose? No such thing: wherever her brother was, there you were sure to find her. A sweeper planted on the lawn became her object of amusement; but her amusements were of the higher order, and to them we stand indebted for the discovery of the comet of 1786, of the comet of 1788, of the comet of 1791, of the comet of 1793, and of the comet of 1795, since rendered familiar to us by the remarkable discovery of Encke. Many also of the nebulæ contained in Sir W. Herschel's catalogues were detected by her during these hours of enjoyment. Indeed, in looking at the joint labours of these extraordinary personages, we scarcely know whether most to admire the intellectual power of the brother, or the unconquerable industry of the sister.

In the year 1797 she presented to the Royal Society a Catalogue of 560 stars taken from Flamsteed's observations, and not inserted in the British Catalogue, together with a collection of errata that should be noticed in the same volume.

Shortly after the death of her brother, Miss Herschel returned to Hanover. Unwilling, however, to relinquish her astronomical labours whilst anything useful presented itself, she undertook and completed the laborious reduction of the places of 2,500 nebulæ, to the 1st of January, 1800, presenting in one view the results of all Sir William Herschel's observations on those bodies, thus bringing to a close half a century spent in astronomical labour.

For this more immediately, and to mark their estimation of services rendered during a whole life to astronomy, your Council resolved to confer on her the distinction of a medal of this Society. The peculiarity of our President's situation, however, and the earnest manner in which the feelings naturally arising from it were urged when the subject was first brought forward, caused your Council to pause,—and waive on that occasion the actual passing their proposed vote. The discussion was, however, renewed on Monday last, and, although there was every disposition to meet the President's wishes, still under a conviction that the actual doing so would have been a dereliction of public duty, it was

Resolved unanimously, "That a Gold Medal of this Society be given to Miss Caroline Herschel, for her recent reduction, to January, 1800, of the Nebulæ discovered by her illustrious brother, which may be considered as the completion of a series of exertions probably unparalleled either in magnitude or importance in the annals of astronomical labour." This vote I am sure every one whom I have the honour to address will most heartily confirm.

Mr. Herschel, in the name of the Astronomical Society of London, I present this medal to your illustrious aunt. In transmitting it to her, assure her that since the foundation of this Society, no one has been adjudged which has been earned by services such as hers. Convey to her our unfeigned regret that she is not resident amongst us; and join to it our wishes, nay our prayers, that as her former days have been glorious, so her future may be happy.*

Extract from the Report of the Council of the Astronomical Society to the Annual Meeting, Feb. 13, 1835. †

" Your Council has no small pleasure in recommending that the names of two ladies, distinguished in different walks of astronomy, be placed on the list of honorary members. On the propriety of such a step, in an astronomical point of view, there can be but one voice; and your Council is of opinion that the time is gone by when either feeling or prejudice, by whichever name it may be proper to call it, should be allowed to interfere with the payment of a well-earned tribute of respect. Your Council has hitherto felt that, whatever might be its own sentiment on the subject, or however able and willing it might be to defend such a measure, it had no right to place the name of a lady in a position the propriety of which might be contested, though upon what it might consider narrow grounds and false principles. But your Council has no fear that such a difference could now take place between any men whose opinion could avail to guide that of society at large ; and, abandoning compliment on the one hand, and false delicacy

---

* The author of this hasty address feels no slight gratification in having been present on the 1st June, 1821, at the last observations with the twenty-foot reflector, in which Miss Herschel was engaged. He remembers also, not without regret, but with becoming gratitude, that the mirror used for his improvement, on the occasion was inserted, for the last time, in the tube, by the hands of Sir William Herschel.—*Memoirs Astronomical Society,* Vol. III., p. 409.

† This extract, as it bears on the subject of the recognition of Miss Herschel's labours, is inserted here, though somewhat before its time.

on the other, submits, that while the tests of astronomical merit should in no case be applied to the works of a woman less severely than to those of a man, the sex of the former should no longer be an obstacle to her receiving any acknowledgment which might be held due to the latter. And your Council therefore recommends this meeting to add to the list of honorary members the names of Miss Caroline Herschel and Mrs. Somerville, of whose astronomical knowledge, and of the utility of the ends to which it has been applied, it is not necessary to recount the proofs. . . . ." *

May 28th, 1828.

DEAR AUNT,—

. . . . Before this reaches you, you will have got it [the medal]. Pray let me be well understood on one point. It was none of my doings. I resisted strenuously. Indeed, being in the situation I actually hold,† I could do no otherwise. The Society have done *well*. I think they might have done *better*, but my voice was neither asked nor listened to.

I ought to mention that it became a matter of discussion at the *Royal* Society whether one of the Royal medals for the year should not be adjudged to you, but the rule limiting the time within which those medals must be granted being precise, it could not be done without a violation of principle.

I have sent by Mr. G. a few copies of a work of mine on Light, for you to distribute. I shall by the next opportunity (*possibly by this*) send some copies of a third catalogue of double stars, completing the first 1,000. The nebulæ are advancing rapidly; I have got about 1,500 reobserved.

Your affectionate nephew,

J. F. W. HERSCHEL.

* " Motions were then made for passing these several resolutions, and the same were carried unanimously."—*Monthly Notices*, vol. iii. p. 91.

† Of President.

MISS HERSCHEL TO J. F. W. HERSCHEL.

June 3, 1828.

MY DEAREST NEPHEW,—

*   *   *   *   *

And I must once more repeat my thanks to you (and perhaps to Mr. South) for thinking so well of me as to exert yourselves for having the great and undeserved and unexpected honour of a medal bestowed on me. . . . .

Here I was interrupted, and all along of the medal; for my friends are all coming to congratulate me, and leave me no time to think of what to say of myself; but I will soon write again, and for the present will only beg that you (or Miss Baldwin, for I dare say she knows,) will give me the history of the medal, such as whose head it is which is on the one side? (I know who it is like very well) and if the impression is to be permanent?

Next, I wish to know if you, or the Royal Society, or the Observatory at Greenwich (the latter I think must be) are in communication with the Imperial astronomer Littrow? If you have seen any of the publications which are yearly printed at the expense of the Emperor, I could wish, if it is not too much trouble to you, to know what you think of the work; because Count Rupfstein, Chargé d'Affaires, sent me the copy (which was to go to Göttingen) to look at, and since then he wants my opinion about it. And I know no more about it than that it is a book printed on fine paper, large folio, of 195 pages, with seven plates of the New Observatory made out of the old one, built at the top of the seventh story of the University at Vienna, a description of the store of instruments, thirty-five articles including rules, two spirit-levels and a case of drawing instruments; tables of precession, aberration, and nutation of ninety-four of the principal stars for the

beginning of the year 1835 ; but I forgot the rest; but so
much I remember, that the whole book is filled with these
ninety-four stars, of which I cannot comprehend the use, but
I say nothing about it, and *hum* and *ha* when the good man
begins to talk about it.   Dear nephew, adieu!
                I am, your affectionate aunt,
                                CAR. HERSCHEL.

   I have but just time to thank my dear Lady Herschel, in
the first place of giving me the great pleasure of seeing her
own handwriting once more, which to me continues much
plainer than all the beautiful new-fashioned Italian hands.
Secondly, I return my best thanks for the beautiful brace·
lets ; I am going to let them be admired this evening, as I
am obliged (though very unwell) to go to a tea-party, and it
will be no small trouble to me to make myself fine enough
for not disgracing your present.
   When next I write I hope I shall not be hurried so, and
be able to tell you how it goes here at Hanover.   Last week
I heard five songs by Madame Catalani at the theatre here ;
but of this, more in my next.
   With many compliments to Miss B.,
        Believe me, your most affectionate sister,
                                C. HERSCHEL.

MISS HERSCHEL TO J. F. W. HERSCHEL.

June 23, 1828.
DEAREST NEPHEW,—
   I have but just time to write a few lines to accompany the Journals Nos. II. and III., therefore I must beg
you to excuse the unconnected manner in which I am
writing, for it must require some time before I, and many a
one beside me, will recover from the fright we were put in
on the 21st, at three o clock in the afternoon, by a thunder-

storm, accompanied with a shower of hail of such an uncommon size as weighing three quarters of a pound ; some
speak of still larger. I, of course, could only judge of
them at a distance by the look, as my carpet was covered by
them of all sizes and shapes ; I noticed one in particular
of the form of a bottle of india-rubber (as it looks before
the neck is cut off), but was at the time incapable of going
near enough, for I was obliged to keep out of the direction
where they entered, forcing the fragments of glass to my
sofa (where I was just going to take my solitary dinner) at
the opposite end of the room, which is twenty-one feet
distant from the window. The houses look deplorable, and
the streets are still glittering with powdered glass. Ex-
presses were sent instantly by the magistrates in all directions to the neighbouring towns and glass-houses for workmen and materials. I have been fortunate enough to get
my lodging-room mended after lying only two nights without
anything but a shutter.

Our gardens and country houses about Hanover have
had the same fate. This happened the day before a *Volks
Fest*, which the Hanoverian Bürgers keep for three days
yearly, and for which all preparations were made, and is
now by many kept with a heavy heart.

But I must not lose this opportunity of mentioning what
I forgot in my last, which is to beg you will (when I am no
more) take my medal under your protection, and give it a
place among those you have of your father's and your own.
I will take care that it shall be delivered to you along with
those books which I keep yet as companions, though it is
seldom I can look into them, for most of my time I am
obliged to waste in lying on the sofa, where I try to forget myself by reading nonsense, over which I soon go to
sleep.

I have the two dullest months before me, for the plays

and concerts do not begin again till autumn ; all families are
either gone to the baths or at their villas, &c. My friends
are all some dozen years younger than myself, and I cannot
always, or but seldom, accept their invitations. Hauptmann Müller took twice tea with me since Christmas. He
heard from Encke that a great astronomical meeting was to
take place at Berlin, to which Mr. South had been invited ;
if there should be any truth in this, and that you and Mr.
South were *inseparables*, I might hope to see you once more ;
but I must not think of anything at the distance, agitations
I cannot bear any longer, I only exist by attempting to be
indifferent about all human events, and hardly anything
can yet give me pleasure but to hear that you, my dear
nephew, and those who are dear to you, are well and happy.

Yours very affectionately,

C. HERSCHEL.

MISS HERSCHEL TO J. F. W. HERSCHEL.

Aug. 21, 1828.

MY DEAR NEPHEW,—

*   *   *   *   *

What you tell me in the short note dated May 24th,
which accompanied the three copies of my Index, concerning the medal, has completely put me out of humour with
the same ; for to say the truth, I felt from the first more
shocked than gratified by that singular distinction, for I
know too well how dangerous it is for women to draw too
much notice on themselves. And the little pleasure I felt
at the receipt of the few lines by your hands, was entirely
owing to the belief that what was done was both with
your approbation and according to your recommendation.
Throughout my long-spent life I have not been used or had
any desire of having public honours bestowed on me ; and

now I have but one wish, that I may take *your* good opinion with me into my grave.

I have no time or inclination to think much on this subject, else I could say a great deal about the *clumsy speech* of the V. P. Whoever says *too much of me* says *too little of your father!* and only can cause me uneasiness.

Mr. South I have seen only twice, or perhaps three times, and that was in yours and your dear father's presence, and to all conversation between you and Mr. South I could only be a listener, and, seeing you so well agree together I congratulated myself on your having found a friend possessing much knowledge of what passes in common life, of which a young and deep mathematician and philosopher has had no time of laying in a great stock.

I heard you would make a visit to Struve at Dorpat this summer together, and I concluded I should then have had a call on the way home. But on that account I feel now relieved from the painful prospect of a final parting from you once more, though it will cost me many melancholy hours to bring that to paper which I yet wish for you to know. But I am too much destroyed at present to explain myself any further, and will only say that by the Michaelmas messenger I will send every scrap of paper which I have yet kept solely for my amusement and for assisting my memory. You may look them over at some leisure [time] and then destroy them; for I go not one night to bed but thinking it may be the last of my life. . . . . I have a numerous and valuable acquaintance, but I keep all my difficulties to myself, for I was ever careful not to injure a relation, or one with whom I am connected, in the opinion of others, by saying what I think of them.

I must prepare to pay a visit at the villa of a friend

of mine where I have twice this summer refused an invitation.

So, God bless you, my dearest nephew, and be assured of my affectionate regard.

C. HERSCHEL.

FROM J. F. W. HERSCHEL TO MISS HERSCHEL.

LONDON, *Dec.* 9, 1828.

MY DEAR AUNT,—

I received your most valuable diary and all the papers you sent me by Mr. Goltermann quite safe, and I most sincerely thank you for them. You speak of "exposing yourself" by presenting them to me, but I am so far from considering it in that light, that I feel proud to possess them, and if anything could increase the regard and esteem I entertain for their writer, it would have been their perusal. Your promised Christmas "scraps and lucubrations" will not be less welcome.

The Journals also came safe and weil to hand, but in the *series* you have sent me I cannot find that for December, 1827, which prevents my binding up the set. If you can procure this and enclose it with the next, I shall be very glad.

I trust to my cousin Mary for telling you all the news of family matters. Astronomy goes on pretty well. My sweeps accumulate. I am very sorry that anything I said should have put you out of humour with the medal, which was a well-merited distinction, and so far as the Astronomical Society is concerned, most honourably conferred. All voices are agreed on that, and on the propriety of the thing, so pray don't suffer yourself to be put out of conceit with it by my nonsense, which after all only went to the manner, not the matter. Our friend S. means well, but wants discretion.

* * * * *

J. F. W. HERSCHEL TO MISS HERSCHEL.

26, Lower Phillimore Place,
Jan. 14, 1829.

My dear Aunt,—

I received your two letters at once, and I cannot enough thank you for the kind consideration which prompted your offer, for I will not yet call it your gift, as I cannot really consent to such a robbery. If you are bent on giving me something truly valuable—infinitely more so than money, which (though I am not rich, and am now less so by some annual hundreds than I was, and am about *voluntarily* * to incur a still further diminution of income) yet, thank God, I am in want of nothing and would rather spare to you than let you spare to me. But if you want to give me what I shall really prize highly, let it be your portrait in oils of the size of my father's. Let me send back the money, and employ part of it in engaging a good Hanoverian artist to paint it. You often tell me your time hangs heavy, so here I am furnishing you with a refuge from *ennui*, and when you know how much pleasure it will give me to see your likeness hanging by my father's, and that you can without inconvenience or difficulty (and *now* without expense) do it, I entreat you not to refuse. I know what you will urge against it, but you undervalue yourself and your own merits so much that I will not allow it any weight.

My mother is ill with the gout, but I hope it is not going to be a severe fit, as she is already on the mend.

Your affectionate nephew,
J. F. W. Herschel.

MISS HERSCHEL TO LADY HERSCHEL.

March 3, 1829.

My dear Lady Herschel,—

I long to congratulate you on the happy occasion of

* An allusion to his approaching marriage, when he would resign his Fellowship.

seeing your dear son so happily settled, but am almost afraid your late illness . . . . may have prevented you from being present at the performance of the ceremony on which the future happiness of my dear nephew is so much depending.

I must beg you will thank Miss B. for sparing me so much of her time by her circumstantial accounts of the interesting event, and hope she will continue to write, though I am not able to answer punctually, for I am not free from pain for one hour out of the twenty-four, and so it has been for a long time past with me. N.B.—She mentions my nephew having written me a letter informing me of his future happiness, but such I have not received, and perhaps he may only have intended it, or it is lost. . . . .

The following hint is only to you as a dear sister, for as such I now know you :—

All I am possessed of is looked upon as their own, when I am gone; the disposal of my brother's picture is even denied me—it hangs in Mrs. H.'s drawing-room, where a set of old women play cards under it on her club day. . . .

I have no great matters to leave, a few articles of furniture which I had the trouble to provide myself with (though I paid for furnished lodgings), would not produce a capital if sold. It is only pictures, books, telescopes, globes, &c., I regret should come into hands of those who know not the value of them ; but Miss Beckedorff will take my sweeper under her protection ; but enough of this. . . . . I hope, above all, to have soon the pleasure to hear that you will hold out with me now that we are entering on our eightieth year.

But as long as God pleases I shall remain

Your most affectionate sister,

C. HERSCHEL.

March 3, 1829.

MY DEAREST NEPHEW,—

I have spent four days in vain endeavours to gain composure enough to give you an idea of the joyful sensation Miss B.'s (and your P.S.) letter of February 5th has caused me. But I can at this present moment find no words which would better express my happiness than those which escaped in exclamation from my lips, according to Simeon. See St. Luke, cap. ii., v. 29 : " Lord, now lettest thou thy servant depart in peace ! "

I have now some hopes of passing the few remainder of my days in as much comfort as the separation from the land where I spent the greatest portion of my life, and from all those which are most dear to me, can admit. For from the description Miss B. has given me of the dear young lady of your choice, I am confident my dear nephew's future happiness is now established.

I beg you will give my love to your dear lady, and best regards to all your new connections where they are due in the best terms you can think of, for I am at present too unwell for writing all I could wish to say.

I have suffered much during this severe winter, and have not been able to leave my habitation above three or four times for the last three months, and feel, moreover, much fatigued by sitting eight times within the last ten days to Professor Tielemann for having my picture taken, which he did at my apartment, and now he has taken it home to finish. You will receive it with the Easter messenger, but I must send it without frame. . . . . I must conclude, for I wish to say a few words to your dear mother. It is now between eleven and twelve, and perhaps you are at this very moment receiving the blessing of Dr. Jennings, in

which I most fervently join by saying, "God bless you
both ! "

Your happy and affectionate aunt,

CAR. HERSCHEL.

TO THE SAME.

March 30, 1829.

DEAREST NEPHEW,—

I have received my picture; by the enclosed card
you will see the name of the artist. . . . . Whatever you
may think about my looking so young, I cannot help; for
two of the days I was sitting to him, I received the agreeable news from England—one day Lady H.'s likeness was
thrown in my lap (Mr. Tielemann taking it out of the box),
and four days after, the account of your approaching happiness arrived. No wonder I became a dozen years younger
all at once. I was sitting about seven hours in so many
days in my own apartments; but there is but one voice,
that the picture looks life itself.

*   *   *   *   *

TO LADY HERSCHEL.

Nov. 16, 1829.

*   *   *   *   *

I was unwilling to be troublesome with a repetition of
the detail of my infirmities, to which I have of late to add
cramps and rheumatic complaints, which rob me of many
hours' sleep and the usual nimbleness in walking, which has
hitherto gained me the admiration of all who know me ;
but the good folks are not aware of the arts I make use of,
which consist in never leaving my rooms in the daytime,
except I am able to trip it along as if nothing were the
matter.

I am glad you are removed again to Kensington, where

you are within a few hours' reach of all who are dear to you (a blessing I never enjoyed throughout all the years of my long life). But I must get in another strain; only when I am writing to you (in particular) I cannot help comparing the country in which I have lived so long, with this in which I must end my days, and which is totally changed since I left it, and not one alive that I knew formerly, except my dear Mrs. Beckedorff; through her means I have, however, been introduced to many valuable ladies of rank and amiable qualities, but to keep up their acquaintance I am obliged to sacrifice my ease and required quiet, which I have still vanity enough to do sometimes.

A fortnight ago I paid my respects to the Landgräfin of Hesse-Homburg (who looks younger and handsomer than when we saw her as a bride at Slough the day before she left Windsor); it was by her desire I made the visit, and I was honoured with a salute at parting, by way of showing we were friends (as she was pleased to say), and a desire to repeat my visit soon.

. . . . I wish also to know on what subject the late Alex. Stewart may have wrote, for that he was an author I know, but I never saw any of his works and might most likely not have understood them, for you know I had no time to read anything for my improvement, but was obliged to be poring for ever over astronomical tables and catalogues, &c.

Another thing I wish Miss B. to inform me of. The 30th November the Royal Society opens with choosing their President and Council; I wish for a list of their names, and likewise of the next change of the Astronomical Society of London. But do not wonder at my being so inquisitive about these things. I cannot think of anything else which could interest me more than to see the names of learned men on paper, especially when I see any of those I have known among them. Besides, as in December our concerts begin, where the Duke of Cambridge, on seeing me, generally makes some inquiry after my nephew and family, and what is going on in the philosophical world, one does not always like to stand with one's mouth open, or to say I cannot tell ! . . . .

Mrs. and Miss Beckedorff send their kind love. . . . .

Mr. Q., 63, he owns himself, marries a young lady in her teens, but she owns 23 ; she could not withstand his pretty equipage. He is grown very old and nasty, and good for nothing but to injure his children and grand-children.

God be with you, my dear Lady H.

Believe me your most affectionate sister,

C. HERSCHEL.

TO J. F. W. HERSCHEL, ESQ.

HANOVER, *January* 11, 1830.

MY DEAREST NEPHEW,—

I am sorry it was not in my power to send a letter by way of announcing the Journals, &c., which you will, I hope, receive soon by the messenger who left Hanover the 27th of December. I have been very ill and confined to my room now three weeks, but it seems der Würg Engel* ist noch einmal vorüber gegangen, at which I am very glad, because I wish to be a little better prepared for making my exit than I am at present.

I intend to amuse myself between this and Easter with collecting and packing up those books which were to be sent to you after my death, and perhaps if I have withstood this *terrible* winter I may have the pleasure of hearing that you have received them safe, and live in the enjoyment of a few months more, in which I hope to hear of the happy increase to your family, and prosperity in general.

So I am to be godmother! with all my heart! I am now

_____
* The Destroying Angel has passed away.

so enured to receiving honours in my old age, that I take
them all upon me without blushing. . . . .

Jan. 12*th.*—No letter for me yet! and no news, excpet
that the Landgräfin of Hesse-Homburg sent me yesterday a
very handsome fur mantle to wear when I go to the play,
with a message that if I did not put it on, by way of saving
it, the next thing she sent me would be a rod. I am accused
of having been clothed too thin, for which I have been suffering these last three weeks. . . . . I will give my opinion,
and in style of a critic, and you will find yourself not to
come off quite free from blame. You have represented me
as a goddess, whereas I have done nothing but what I
believe to be right; and wherever I did wrong, it was
because I knew no better!

MISS HERSCHEL TO J. F. W. HERSCHEL, ESQ.

HANOVER, *June* 18, 1830.

MY DEAREST NEPHEW,—

This letter will go by to-day's post, which I believe
is the last before the messenger leaves Hanover, and Lega-
tions Rath Haase has promised to direct the box for me, so
that it is to be called for at Mr. Goltermann's either by
yourself, or somebody who will look to it, that it may come
safely to your hands. And I will give you here a list of
the contents of the box, by which you will see that I must
be very anxious till I know that it is safely come to hand,
especially as I was obliged to have the box made very slight
on account of saving size and weight.

Contents :—

Wollaston's Catalogue.

Bode's Catalogue.

My Index to Flamsteed's Observations.

Herschel's and South's Observations, bound in red
morocco.

Logarithmic Tables by Taylor.

Seventy-two Papers of your father's, in five volumes.

The parcel directed for my niece contains ornaments which I am afraid will soon be wanted for a general mourning, but I am told they may be worn at any time. Miss Beckedorff chose them for me; my direction was they should be pretty, and not of English manufacture, and not larger than what might be put in the space which I showed her. I am only sorry I could not find anything that might please your dear mother, for, to judge by myself, we want now only ease, quiet, and patience to bear the pains and infirmities attendant on our age; and we are too far asunder for doing more than wishing one another the above-mentioned qualities.

*   *   *   *   *

I had intended to have sent my medal along with the books, but since you have presented me with the handsome miniature of your dear Margaretta, from which I cannot part as long as I live, I have mentioned already to Dr. Groskopff that the medal, miniature, and my gold watch [the gift of her grandfather in 1774], are to be sent to my grand-niece and namesake, C. H.

*   *   *   *   *

I do not like to send empty paper, but I must. Time falls short, and I am tired already with the thought of the long walk I have to take to carry this letter, for I must see Haase once more, and it is attended with great difficulty to get so heavy a box over at present.

God bless you, dear nephew,

Says your affectionate aunt,

CAR. HERSCHEL.

MISS HERSCHEL TO J. F. W. HERSCHEL, ESQ.

HANOVER, *Oct.* 27, 1830.

MY DEAREST NEPHEW,—

I see by my memorandum-book that I sent a letter to your dear mother on the 20th August, partly in answer to one of Miss Baldwin's, which contained the melancholy account of Miss Isabella [Stewart's] dangerous state of health. I have ever since been very uneasy, and wishing for more cheering information, because I know what a drawback it would be on the happiness of all your dear connection if you should lose her, besides the interruption it must cause in the hitherto cheerful correspondence in which even my dear niece took the pen to join in affording me the only comfort I am yet capable of receiving. . . . .

Tell your dear Margaret that the very day on which the letter arrived, in which she requested some hair, I sent for the hair-dresser and made him cut off all which was useless to me, leaving plenty for a toupee and a little curl in the poll. But I repented not having kept a few out of the plait, which I might have sent in a letter, as I understand it is designed for a talisman against the evils of this hurly-burlying world. But I consoled myself with the thoughts that no harm could possibly assail the dear little creature as long as she is under the care of her affectionate and excellent mother, leaving a loving father out of the account.

Dr. Groskopff has been zum Ritter ernannt by his present Majesty. So was Dr. Mükry last week. If all is betitled in England and Germany, why is not my nephew, J. H., a lord, or a wycount at least (query)? General Komarzewsky used to say to your father, Why does not he (meaning King George III.) make you Duke of Slough?

      *   *   *   *   *

MISS HERSCHEL TO J. F. W. HERSCHEL, ESQ.

March, 1831.

MY DEAREST NEPHEW,—

If it was not high time to congratulate you on your birthday—of which I most heartily wish you may enjoy many returns in uninterrupted and increasing happiness—I might have still deferred to thank you for your kind letter and the valuable present of *your book.* I intend to follow your mother's example to read it "from end to end," which I was hitherto not able to do on account of my dim eyes ; but now the days are getting longer I think I see better, and to judge by the few pages I have read, that so far from making me go to sleep, it will be an antidote against a propensity for doing so in the daytime.

I much regret my inability to acknowledge my dear niece's letter in such a manner as might encourage a corre-spondence with me, but it is difficult to write in a cheerful strain when one is continually in the dismals. I do all I can to keep up my spirits under a daily increase of my in-firmities, and have been best part of the winter confined to my rooms. My complaint is incurable, for it is a *decay* of nature, and nine days after your birthday I am eighty-one. What a shocking idea it is to be decaying ! *decaying !* But never mind—if I am decaying here, there will be, as Mrs. Maskelyne once was comforting me (on observing my grow-ing lean), "the less corruption in my grave ! "

22*nd.*—Some weeks ago I wrote as above, which I in-tended as a preface to my dying speech, with intention to give you a few hints concerning ——, and indeed I may say of all my German relations, except the Knipping family. If I did not fear that some of them would, after my decease, introduce themselves as troublesome correspondents to you, I would rather write about something else just now, and indeed I had better drop the subject, for you will know, I suppose, how to rid yourself of a pestering fool by answering coolly, or not at all.

23rd, afternoon.—Yesterday I was interrupted again, and the whole morning of the present, which I had intended to spend with you at Slough, has again been taken up with gabbling with my radical servant. But the day after Easter I get another, and I hope I shall have better luck; but till then I am not mistress of my time, therefore will hasten to inform you that Mrs. Beckedorff is packing up a parcel for me, which is going from here the day after Easter. . . . . The packet contains a tablecloth, with twelve napkins (the cloth is eight yards long, Mrs. B. says), which I hope my dear niece will do me the pleasure to accept as a remembrance of her old aunt.

Your book * I have read as far as page 150, and met with nothing but what I clearly can comprehend, and promise myself much pleasure in reading the rest, which hitherto I have been prevented to do by being continually interrupted, and besides not being able to read many pages at a time before the lines run one into another.

My dear niece said in her letter to me your book would cause a sensation, and so it has, as I hear from all quarters. I am told it has been translated into German from a French translation, and much [all in admiration] is appearing in Gelehrten Anzergen, which I have not yet been able to get a sight of. . . . . I must give over and defer writing till I am provided with pen, ink, and paper. The first thing my radical servant did when she came to me was to break the bottle [containing] the ink of my own making, which was to have lasted me all my life-time. . . . . First and foremost, give my love to your dear mother, and believe me, ever your most affectionate aunt,

C. HERSCHEL.

* Discourse on the study of Natural Philosophy.

MISS HERSCHEL TO MRS. HERSCHEL.

HANOVER, *May* 14, 1831.

O ! my dearest niece, where shall I find words which can express my thanks to you for writing me such an interesting letter, at a moment when you were suffering from indisposition !

*   *   *   *   *

May 18*th.*—Dear niece, how are you now ?   I hope so far well enough to read what I think necessary to say in answer to yours of May 2nd.   I was glad to see that you think the table-linen pretty, but I tremble on seeing that you puzzle yourself about sending me anything in return.   Nothing would distress me more than receiving anything from England besides such dear letters as I have hitherto been blessed with, for I am provided with even more than is necessary to appear in the best circle of society, whenever my feebleness will permit me to go from home, and I feel no small regret at leaving so many good things among those who do not want it, or ever cared for me.   Now, this is once for all ! and you have nothing to do but to go by what dear Herschel says—he knows me, I see, better than I thought he did.

I have something to remark about what you call my letters, which were to be deposited in the letter case.   I was in hopes you would have thrown away such incoherent stuff, as I generally write in a hurry at those moments when I am sick for want of knowing how it looks at home, and not to let it rise in judgment against my, perhaps, bad grammar, bad spelling, &c., for to the very last I must feel myself walking on uncertain grounds, having been obliged to learn too much without any one thing thoroughly ; for my dear brother William was my only teacher, and we began generally with what we should have ended ; he, supposing

I knew all that went before. Perhaps I might have done so once, but my memory he used to compare with sand, in which everything could be inscribed with ease, but as easily effaced. Some time hence you will see a book* in which I transcribed such lessons as my brother was obliged to give me at such times when I was to set about some calculations of which I knew not much about. I shall this summer collect every scrap of that kind—some written by my brother, and some penned down as they flowed from his lips, and some even incomplete, which were intended to be given more correct when at leisure. I bought a very handsome portfolio for this purpose, and had my nephew's new seal engraved upon the lock.

I should not have thought of troubling my dear nephew or you with looking over these fragments, but I cannot part with remembrances of times long gone by, so long as life is in me; but for fear I should not have at the last moment the power of burning them, I will keep them ready for being sent off to Slough, for nothing of the kind shall be seen by unhallowed eyes. . . . .

MISS HERSCHEL TO J. F. W. HERSCHEL.

HANOVER, *June* 4, 1831.

MY DEAREST NEPHEW,—

Just now I received yours of May 22nd, and the next post will not go from here till the 7th, and I wish the wind may be favourable that you may be soon made easy about the £50, for which I beg you will, according to custom, give the above receipt to your dear mother. And you may as well add my heartfelt thanks; for what good can it do troubling her with my letters, knowing the weakness in her hands will not permit her answering them. . . . .

* See p. 72, 1786.

. . . . I have laid apart for every possible expense which can occur at my exit. Six years ago I had a vault built in the spot where my parents rest. The ground is mine auf ewig (for ever).

You have made me completely happy for some time with the account you sent me of the double stars; but it vexes me more and more that in this abominable city there is no one who is capable of partaking in the joy I feel on this revival of your father's name. His observations on double stars were from first to last the most interesting subject; he never lost sight of it in his papers on the construction of the heavens, &c. And I cannot help lamenting that he could not take to his grave with him the satisfaction I feel at present at seeing his *son* doing him so ample justice by endeavouring to perfect what he could only begin. . . . .

TO MRS. HERSCHEL.

HANOVER, *August* 11, 1831.

. . . . I wish Paganini may make some stay yet in England, that you, or my nephew at least, may hear him. The English cannot be more frantic about him than the Hanoverians were. He filled our play-house twice at double price, and though some part of the orchestra had been thrown into parquet, still gentlemen were scattered among the lamps and squeezed in among the performers on the stage. You will think me the maddest of the mad when I tell you that, after spending three parts of each day in pain and misery, I make one of the audience twice a week, if possibly I can hold up my head ; for then I am lulled into forgetfulness of my severed situation from all what *was* or *is* still dear to me, and amuse myself sometimes with having my vanity tickled by the notice which is taken of my being or not being present. In *The Sun* of July 13 is a description of Paganini's face and looks, which I could not have given better myself after having had some conversation with him (through an interpreter) ; on coming one evening at the end of a play out of my box, I found some gentlemen waiting to introduce him to me, which I believe was partly done to give the people an opportunity to see him.

I am reading all the Parliamentary speeches as given in *The Sun,* and there I meet with some excellent ones by a Sir James Mackintosh ; pray is he any connexion of your family ? In the paper of July 6th I saw a quotation (by a speaker, Mr. E. Lytton Bulwer,) from a celebrated philosopher (meaning our *own* J. Herschel) who had felicitously observed that " the greatest discoverer in science can do no more than accelerate the progress of discovery." . . . .

I remain, my dear niece,

Your most affectionate

CAR. HERSCHEL.

The following letter, from the celebrated Encke, is one of the few preserved which belong to this period, and gives graceful expression to the high esteem in which she was held :—

FROM PROFESSOR ENCKE TO MISS HERSCHEL.

BERLIN, *Aug.* 17, 1831.

MADAME,—

I feel great pleasure in informing you that the parcel which has been forwarded to me through your kindness is safely arrived here, and has been delivered to Professor Mitscherlich, according to the directions given by your celebrated nephew, J. Herschel.

I hardly know, madame, how to return you my thanks for the trouble you have so kindly taken in transmitting the parcel to me. It would, indeed, have been an irretrievable loss to have been deprived of the excellent treatise written by your eminent nephew, had it not reached its destination.

Allow me, madame, to avail myself of this opportunity to pay my respects to a lady, whose name is so intimately connected with the most brilliant astronomical discoveries of the age, and whose claims to the gratitude of every astronomer will be as conspicuous as your own exertions for extending the boundaries of our knowledge, and for assisting to develope the discoveries by which the name of your great brother has been rendered so famous throughout the literary world.

I am, with great esteem and regard, madame,
Your most obedient, humble servant,
T. F. ENCKE.

MISS HERSCHEL TO SIR J. F. W. HERSCHEL.

HANOVER, *Oct.* 25, 1831.

MY DEAR SIR JOHN,—

But mind, you are still my dear nephew, and will be so good as to give the above to your dear mother. With this last sum, I have actually received since I am here a thousand pounds; a sum which I had no idea (nor I am sure your father neither) you would have been burdened with so long, for when I left England I thought my life was not worth a farthing. But no more of this for the present. . . .

You promised me another Catalogue of double stars, but I suppose you have had no time to arrange them. But do not observe too much in cold weather. Write rather books to make folks stare at your profound knowledge. . . . . .

Loves and compliments to all whom *we* love, and God bless my dear nephew, says
Your affectionate aunt,
CAR. HERSCHEL.

P.S.—I received Miss B.'s letter on the 16th. It gave

me infinite pleasure to see that Babbage and Brewster have
also been honoured with notice. As for the news of my dear
nephew's appointment, she came too late, for on the 9th I
was honoured by a note written by the Duke of Cambridge's
own hands, informing me of it.

MISS HERSCHEL TO SIR JOHN HERSCHEL.

HANOVER, *December* 25, 1831.
DEAREST NEPHEW,—
    More than two months are elapsed since I was made
happy by your dear letter of October 15th . . . . I hope
that perhaps some good account is on its passage and may
reach me before the rivers are frozen up, as at this time of
the year the posts are often interrupted.

    I have of late been very little from home, except two
evenings in the week to the play, for I cannot walk the
streets without being led, as I cannot trust my eyes to avoid
obstacles, besides a total loss of strength ; so that the chief
connection I keep up with this world depends on what I by
imperfect glimpses can gather from the newspaper and a
little talk sometimes with Mrs. Beckedorff. But a few
weeks ago I exerted myself, fearing if I delayed much
longer I might not be able at all to pay my respects to our
good Duchess of Cambridge, and I wished to make good a
blunder I had committed two years ago, when I was conversing with her at the Landgräfin's for half an hour
together, taking her all the while to be an officer's lady, as
she came accompanied by her brother, the Prince of Hesse,
who wore a moustache. It is the case in general, that I
do not know my most intimate friends except by their
voices. I was, however, very much gratified by my visit.
A lady, who is in the habit of going to Court, left my name
along with her own with the lady-in-waiting, and the next
Sunday we were appointed to be there at half-past one (a

very inconvenient hour for me, for I only begin to be alive
when other folks go to sleep). But no reception could be more
friendly. I was made take my place by her on the sofa, and
after some conversation, the little Princess Augusta was
called to tell me that she had seen you at Slough ; you had
shown her the telescope and described how it was moved by
the handle round about. I asked her if she had seen the
little girls. The Duchess explained that her call had been
unexpected, and regretted that she had not had an opportunity of coming to Slough herself. Then the Princess was
sent to call her father, whom I presented with your book,
and he went to fetch his spectacles, and was much pleased
with the subject, saying, " I shall read it, for I like such
things." After I had read the whole book myself—mind, I
say *the whole*, though you recommended me to read only
the first and last chapters—and knowing no one who is
worthy to look into it, I had it handsomely bound and
wrote in the top margin "To His Royal Highness, the
Duke of Cambridge." At the side of Sir Francis Bacon
stands "*from*" and in the margin at the bottom, " Caroline
Herschel, aunt of the author." By this means, I know it
secured from contamination in the Duke's library, where
anybody who is desirous of reading it will find it.

December 26th.

My dearest Niece,—

So far I wrote last night, thinking to fill this page
to-day, with such news as I should like to communicate to
my nephew if he was present ; but now all is fled from my
memory, for my dear sister is ill, and perhaps still in danger,
and my only trust is in your goodness of sending me a
speedy account, which may confirm the hope you seem to
entertain of her recovery. For there is nothing I so
ardently desire as to be spared the pain of mourning for a

single individual of those friends I have in England, and
how much more it would affect me to lose one so nearly
connected, and within a few months of my own age, it may
be easily imagined . . . . Next to listening to the conversation of learned men, I like to hear about them, but I find
myself, unfortunately, among beings who like nothing but
smoking, big talk on politics, wars, and such like things.
Of our German astronomers, I have lately heard nothing;
but that, perhaps, is owing to Encke having had the cholera,
but of which he soon recovered. Gauss has been long unhappily situated by losing his second wife, who had been
long lingering . . . .

. . . . I beg once more for an early assurance of my dear
sister's recovery.

MISS HERSCHEL TO SIR JOHN HERSCHEL.

HANOVER, *Jan.* 20, 1832.

MY DEAREST NEPHEW,—

My dear niece's and your letter of January 3rd, have
indeed answered your kind intentions, for the painful communication of your last found me prepared, and enabled me
to break the black seal with tolerable composure, and I
found no small consolation from your description of the easy
ending of your dear departed parent.

At this moment, I am incapable of saying anything of
myself. I know it cannot be long before I shall follow the
dear departed, and my pen would trace nothing but lamentations at the prospect that my remains will not be joined
in rest by the side of those with whom I lived so long.

But I beg and trust you will continue to bless me with
your good opinion and approbation, until the close, for that
I have hitherto been in possession of the same, I conclude
from the kind letters I receive from your own hands. . . . .

MISS HERSCHEL TO LADY HERSCHEL.

HANOVER, *March* 14, 1832.

MY DEAREST NIECE,—

Your precious letter, which I received this morning, has relieved my mind from the fear that some ill might have befallen my dear friends, because in my solitude the time between January 7th and March 14th, seems to be an age ; besides, the last melancholy letters required some soothing subject to think on, for I do nothing else but think of the spot where I once was and never can be again.

But now all is well ; your dear letter will make me happy for some time to come, and in my next I will more fully reply to it, when I hope to be more composed than I am just now, as the day after to-morrow will be my birthday, when I, *perhaps*, enter on my eighty-third year. I am always at the return of that day what one may call "hipt," and therefore must destroy my thoughts any how as well as I can.

I kept my dear nephew's birthday last week, the 7th of March, by thinking of you throughout the whole day. When I was at dinner, I made my maid stand opposite to me, and pouring her out a glass of wine, made her say, Sir John Herschel, lebe hoch ! (for ever).

But I must hasten to say that which I wish you to know as soon as possible, which is, to beg of all things not to send the parcel the good Miss B. intended for me. I suppose it may consist of some dress of my dear departed sister . . . . I beg your acceptance of it for a remembrance of us both ; it would vex me to add anything I set store on, only to leave it to those I cannot esteem.

*   *   *   *   *

I am much obliged to my dear nephew for sending the few pages announcing the publications of the Royal Society. It is only such morsels as these which keep up a desire for living any longer. But the premium of the King of Denmark's medal, for the discovery of telescopic comets, provokes me beyond all endurance, for it is of no use to me. One of my eyes is nearly dark, and I can hardly find the line again I have just been tracing by feeling on paper.

Pray do not forget me when my nephew's recension of Mrs. Somerville's works makes its appearance. . . . .

TO SIR J. F. W. HERSCHEL.

HANOVER, *April* 20, 1832.

MY DEAREST NEPHEW,—

*   *   *   *   *

My dear niece has promised me your article * on the writings of Mrs. Somerville. I hope she will not forget it, nor you the Catalogue of double stars. Such things make me very happy, but of any expensive publications I would not wish you to throw away upon me *now;* it makes me only grudge to think of having to leave them in the hands of blockheads. But if you have anything for Göttingen, Encke, or Bessel, it amuses me to forward it. Olbers has been dangerously ill for some time; they tell me he is too fat, and lives too well.

I only write this by way of announcing the parcel, that you may inquire for it should it not come to hand in due time, else I am very tired, and must yet make up the parcel, and I want to show myself once more to-morrow evening at the Oratorio, as it is for the poor, and will be the last performance this season. . . . .

EXTRACT FROM A LETTER OF SIR JOHN HERSCHEL.

HANOVER, *June* 19, 1832.

. . . . I found my aunt wonderfully well and very nicely and comfortably lodged, and we have since been on the full

* In the Quarterly Review.

trot. She runs about the town with me and skips up her
two flights of stairs as light and fresh at least as *some folks*
I could name who are not a fourth part of her age. . . . . .
In the morning till eleven or twelve she is dull and weary,
but as the day advances she gains life, and is quite "fresh
and funny" at ten or eleven, p.m., and sings old rhymes, nay,
even dances! to the great delight of all who see her. . . . .
. . . . It was only this evening that, escaping from a
party at Mrs. Beckedorff's, I was able to indulge in what
my soul has been yearning for ever since I came here—a
solitary ramble out of town, among the meadows which
border the Leine-strom, from which the old, tall, sombre-
looking Markt-thurm and the three beautiful lanthorn-
steeples of Hanover are seen as in the little picture I have
often looked at with a sort of mysterious wonder when a
boy as that strange place in foreign parts that my father
and uncle used to talk so much about, and so familiarly.
The *likeness* is correct, and I soon found the point of view.

Yesterday, being the anniversary of Waterloo, there was
a great military *spectacle* here in a large esplanade, where
there is erected a tall and very pretty column, with a
bronze "Victory" at the top, hopping on one leg. A few
guns were fired, a sermon preached, the veil of the statue
(shown for the first time) pulled off by the Duke of Cambridge, and a good dinner eaten by 350 personages, of
which number I had the honour to be one unit, in a vast
saloon in the Herrenhauser Palace, about the length,
breadth, and height of St. George's Hall, at Windsor, the
Duke presiding and giving the toasts, &c., in honour of
the Waterloo heroes. The saloon was ornamented most
curiously with guns, swords, and pikes, arranged in patterns,
and with Waterloo trophies, and a panoramic view of the field
of Waterloo in compartments. No ladies were admitted to
the table, and (what say you to the gallantry of the Hano-

verian military ?) there was no ball in the evening, nor *any*
the *slightest* provision for the amusement or participation of
the fair. So Mars and Venus, I suppose, have had a
" tiff ! "

ADIEU.

TO LADY HERSCHEL.

HANOVER, *Dec.* 4, 1832.

MY DEAREST NIECE,—

I shall in future, when I have anything to say to my
dear nephew address myself to you, well knowing his time
is too precious for spending even on reading. . . . .
Thank him most heartily for the " Edinburgh Review," and
the description of the wonderful machine. . . . . But here
is the *grievance*—I cannot possibly read the Review, my
sight is almost lost, and I must wait till Miss Beckedorff or
somebody can read to me. . . . . Dr. Tias, who travelled
through Hanover, called on me to day. He talked strangely
about my nephew's intention of going to the Cape of Good
Hope. Mr. Hausmann told me some weeks ago that the
Times contained the same report, to which I replied, " It is
a lie ! " but what I heard from Dr. Tias to-day makes me
almost believe it possible. Ja ! if I was thirty or forty
years younger, and could go too ? in Gottes nahmen ! But
I will not think about it till you yourself tell me more of it,
for I have enough to think of my cramps, blindness, sleepless nights, &c.

TO SIR J. F. W. HERSCHEL.

HANOVER, *March* 30, 1833.

MY DEAREST NEPHEW,—

Ever since the 6th of March, the day on which I
received my dear niece's of the 26th of February, I have

been enabled to dispel by its comfortable contents the
gloomy reflections with which I am on the return of your
and my birthdays assailed. But being obliged to spend
such days alone, at a distance from all who are dear to us ;
or, what would be worse, in the presence of beings of un-
congenial feelings, one is apt to fall again into the dismals,
which the return of the late snow and frosty weather prevented my taking recourse to my usual remedy, which is to
turn all grievances into a joke. Your birthday I celebrated
exactly like that of 1832, viz., after dinner I jingled glasses
with Betty, and made her say, " Es lebe Sir John ! hoch !*
hurrah ! " She went in the kitchen to wash the dishes, and
I with a book (a silly novel) in my hand on the sofa asleep !
. . . .

I begin to be confused, and had rather say nothing of the
thousand things which are running in my head, and which
all must be said within the next six months. As yet I can
follow your steps and proceedings, for I read the papers—the
Globe—and saw that in June is the meeting in Cambridge.
. . . . From these papers I also see how all my valuable
acquaintances drop off one after another. Captain Kater
has lost his wife, the fine singer ; Mrs. Parry ; Lady Harcourt ; your dear mother, are gone—the latter three of my
own age, and I must hold out !

TO LADY HERSCHEL.

HANOVER, *August* 1, 1833.

*   *   *   *   *

I have now the pleasure of thanking my nephew for
his valuable book of astronomy, having actually received it
by yesterday's post, and by a kind letter from Professor
Schumacher. I learn that I may yet hope to see the
promised Catalogue of nebulæ and double stars, to the

* Sir John for ever !

perusal of which I look forward as a solace during the time you will be on your way *far, far* from us. But these treasures cause me no little thinking about in whose hands I shall leave them when I cannot see them any longer, but cannot think of anyone I should like to leave them in preference to the Duke of Cambridge.

I cannot find words which would express sufficient thanks to my dear nephew for his last letter, every line of which conveys a comfort.

    *   *   *   *   *

P.S.—Dear Nephew, as soon as your instrument is erected I wish you would see if there was not something remarkable in the lower part of the Scorpion to be found, for I remember your father returned several nights and years to the same spot, but could not satisfy himself about the uncommon appearance of that part of the heavens. It was something more than a total absence of stars (I believe). But you will have seen by the register that those lower parts could only be marked *half swept.* I wish you health and good success to all you undertake, and a happy return to a peaceful home in old England. God bless you all!

TO THE SAME.

Sept. 6, 1833.

My dear Niece,—

Eight days are already gone since the arrival of your dear letter of August 21st, and I can hardly muster up composure enough at this moment to reply to it, because my ideas are still, what they ever have been, more occupied with future or past events than what passes immediately about me. At present my thoughts are wholly fixed on the busy scenes with which you are at present surrounded, and regretting that I am not with you to afford you any assis-

tance, or to take charge of my nephew's workshops, as I
used to do of his father's when absent; or that it is not
possible to shake off some thirty years from my shoulders
that 1 might accompany you on your voyage.

In answer to your query about my nephew's building a
grotto of coals I must plead ignorance, but have no doubt
many an edifice of that kind has daily been erected and
erased without my being present, for my dear nephew was
only in his sixth year when I came to be detached from the
family circle. But this did not hinder *John* and *I* from remaining the most affectionate friends, and many a half or
whole holiday he was allowed to spend with me, was dedicated to making experiments in chemistry, where generally
all boxes, tops of tea-canisters, pepper-boxes, teacups, &c.,
served for the necessary vessels, and the sand-tub furnished
the matter to be analysed. I only had to take care to ex-
clude water, which would have produced havoc on my carpet. And for his first notion of building I believe he is
indebted to me, for it was on his second or third birthday
when I lifted him in the trenches to lay the south corner-
stone of the building which was added to the original house
at Slough. It must have been the second year of his age,
for I remember I was obliged to use a deal of coaxing to
make him part with the money he was to lay on the brick.

About the same time, when one day I was sitting beside
him, listening to his prattle, my attention was drawn by
his hammering to see what he might be about, and found
that it was the continuation of many days' labour, and that
the ground about the corner of the house was undermined,
the corner-stone entirely away, and he was hard at work
going on with the next. I gave the alarm, and old John
Wiltshire, a favourite carpenter, came running, crying out,
" God bless the boy, if he is not going to pull the house
down ! ' (Our John was this man's pet, he taught him to

s 2

handle the tools). A bricklayer came directly with brick and mortar to mend the damage.

I was called to my solitary dinner just when I was going to give you a few specimens of my nephew's poetry; I have some by me, composed when about eight or nine years old, in a most shocking handwriting; but generally about this time I am so sleepy that I think it will be best to give you the continuation in a posthumous letter from C. H. to Lady M. B. Herschel, to be delivered to her on her return from the Cape. . . . .

If I only live long enough to have the assurance of your *all* being well and safely got to the Cape, I will lay down my head in peace.

My paper is not filled, but there is not time for writing more, nor do I like to think about the present; but about a month ago I began a day-book again, which I was in the habit of keeping when in England, and with the contents of that I intend to fill my posthumous letter to you.

God bless you, my dear niece . . . . and with my love to my dear nephew and yourself,

I remain,

Your most affectionate aunt,

CAR. HERSCHEL.

TO PROFESSOR SCHUMACHER.

Dec. 11, 1833.

DEAR SIR,—

By recollecting your former obliging kindness to me, I am encouraged once more to intrude on your valuable time by transcribing part of my nephew's last letter, dated from Portsmouth, November 10th:—" The last proof sheet of my nebulæ paper left my hands the night I left London, and yesterday I got twelve copies to take to the Cape. One will be forwarded to you to-morrow by Lieut. Stratford,

R.N., superintendent of the "Nautical Almanac," who will send it to Prof. Schumacher, to whom, if you do not soon get it, pray write. I have also ordered a duplicate to be sent you by Mr. Hudson, assistant secretary of the Royal Society, and librarian, who will henceforward send you all my papers (in duplicate). My observations on the satellites of Uranus, which confirm my brother's results, were sent to be put in course of publication last night."

I have no doubt but that you, Sir, are in correspondence with the above named, but to me unknown, gentlemen, and that those two copies intended for me are only enclosed in a packet with many for yourself.

I long much to see the observations on the Georgian satellites, but doubt their being ready to come with the paper on nebulæ. I beg you will order them to be forwarded to me as soon as you see them yourselves, for I do not flatter myself with the hopes of being much longer for this world, but will be thankful if life is spared me till the end of April, when I hope to receive the assurance of my nephew's safe arrival with his dear family at the Cape.

Excuse my troubling you so far, and believe me with great regard, dear Sir,

Your much obliged and humble servant,

C. HERSCHEL.

CHAPTER VII.

CAPE TOWN, *Jan.* 21, 1834.

MY DEAR AUNT,—

Here we are safely landed and comfortably housed at the far end of Africa, and having secured the landing and final stowage of all the telescopes and other matters, as far as I can see, without the slightest injury, I lose no time in reporting to you our good success *so far*. M. and the children are, thank God, quite well; though, for fear you should think her too good a sailor, I ought to add that she continued sea-sick, at intervals, during the whole passage. We were nine weeks and two days at sea, during which period we experienced only one day of contrary wind. We had a brisk breeze " right aft " all the way from the Bay of Biscay (which we never entered) to the " calm latitudes," that is to say, to the space about five or six degrees broad near the equator, where the trade winds cease, and where it is no unusual thing for a ship to lie becalmed for a month or six weeks, frying under a vertical sun. Such, however, was not our fate. We were detained only three or four days by the calms usual in that zone, but never *quite* still, or driven out of our course, and immediately on crossing " the line," got a good breeze (the south-east trade wind), which carried us round Trinidad, then exchanged it for a north-west wind, which, with the exception of one day's squall from the south-east, carried us straight into Table Bay. On the night of the 14th we were told to prepare to see the Table Mountain. Next morning (N.B., we had not seen

land before since leaving England), at dawn the welcome
word "land" was heard, and there stood this magnificent
hill, with all its attendant mountain range down to the
farthest point of South Africa, full in view, with a clear blue
ghost-like outline, and that night we cast anchor within the
Bay. Next morning early we landed under escort of Dr.
Stewart, M.'s brother, and you may imagine the meeting.
We took up our quarters at a most comfortable lodging-
house (Miss Rabe's), and I proceeded, without loss of time,
to unship the instruments. This was no trifling operation,
as they filled (with the rest of our luggage) fifteen large
boats; and, owing to the difficulty of getting them up from
the "hold" of the ship, required several days to complete
the landing. During the whole time (and indeed up to this
moment) not a single south-east gale, the summer torment
of this harbour, has occurred. This is a thing almost un-
heard of here, and has indeed been most fortunate, since
otherwise it is not at all unlikely that some of the boats,
laden as they were to the water's edge, might have been lost,
and the whole business crippled.

For the last two or three days we have been looking at
houses, and have all but agreed for one, a most beautiful
place within four or five miles out of town, called "The
Grove." In point of situation, it is a perfect paradise, in
rich and magnificent mountain scenery, and sheltered from
all winds, even the fierce south-easter, by thick surrounding
woods. I must reserve for my next all description of the
gorgeous display of flowers which adorns this splendid
country, as well as of the astonishing brilliancy of the constellations, which the calm, clear nights show off to great
advantage; and wishing we had you here to see them, must
conclude with best loves from M. and the children.

Your affectionate nephew,

J. F. W. HERSCHEL.

BRAUNSCHWEIGER STRASSE, No. 376,
May 1, 1834.

MY DEAR NEPHEW,—

Your precious letter relieved me on the 14th from a whole twelvemonth's anxiety, for it was in April last year when, by your few brief lines on business, I saw that you were seriously preparing for leaving Europe, and from that time I became in idea a vagrant accompanying you through all the fatigues of preparing for such a momentous under-taking. And if it had not been for the consoling letter of your brother [in law] James, and one from Miss B. giving me an account of the carefully arranged accommodation with which they saw you depart, I should not have known how to support myself till I saw your dear letter, which brought me even more comfort than I could hope you would have found time to think of. . . . .

Both yourself and my dear niece urged me to write often, and to write always twice; but alas! I could not overcome the reluctance I felt of telling you that it is over with me, for getting up at eight or nine o'clock, dressing myself, eating my dinner alone without an appetite, falling asleep over a novel (I am obliged to lay down to recover the fatigue of the morning's exertions) awaking with nothing but the prospect of the trouble of getting to bed, where very seldom I get above two hours' sleep. It is enough to make a parson swear! To this I must add I found full employment for the few moments, when I could rouse myself from a melancholy lethargy, to spend in looking over my store of astronomical and other memorandums of upwards of fifty years collecting, and destroying all what might produce nonsense when coming through the hands of a Block-kopff in the Zeitungen.

My dear friends, Mrs. and Miss Beckedorff, are assisting

me in my final preparations for going to that bourn from whence none ever returned, but let me hope that you, my dear nephew, with my dear niece and the whole of your young family, will return to your dear relatives and friends after having seen all your wishes and expectations crowned with success. Though, if I may not be among those who will greet your return, I can assure you their number will be *great*, judging from the sensation the account of your safe arrival at the Cape has caused among all our friends; and (as Dr. M—— will have it) " the whole intelligent and scientific world in general are participating in our feeling." Poor Mrs. Beckedorff, to whom I read your letter, sat trembling and crying for joy; for I now find that my friends had not been without fear for your safety on account of the storms (and their sad consequences) which prevailed for a long time immediately after your departure, and the same evening a note was despatched to her Royal Highness the Landgräfin to communicate the news; for from the Duke's and her Royal Highness's constant inquiries when I expected to hear from you, I knew the account of your safe arrival would give pleasure.

*   *   *   *   *

The feelings of joy I experienced the first few days after the arrival of your letter are nearly evaporated, and I begin to feel already that the essential information required for making me reconciled to the immense space which divides me from you is still wanting; which is, that I cannot now, as formerly, receive so frequent accounts concerning the health of my dear niece and the children, not even from Miss B., who used to describe their little ways so prettily, for she, too, cannot now observe them. I look with impatience for the next account . . . . of the health of my dear niece, yours, and the dear little beings. Caroline and Isabella and I are old friends, but is William Herschel the second likely to live (if not beyond) at least to the age of his grandfather?

Perhaps you will receive the " Göttingsche Gelehrte Anzeigen" of 16th and 19th December, 1833, containing what is said of your book on Natural Philosophy (by Gauss they say).

God bless you *all*, and believe. my dear nephew,

Ever your most affectionate aunt,

CAR. HERSCHEL.

FROM J. F. W. HERSCHEL TO MISS HERSCHEL.

FELDHAUSEN, *June* 6, 1834.

MY DEAR AUNT,—

The twenty-foot has been in activity ever since the end of February, and, as I have now got the polishing apparatus erected and three mirrors (one of which I mean to keep constantly polishing) the sweeping gets on rapidly. I had hardly begun regular sweeping, when I discovered two beautiful planetary nebulæ, exactly like planets, and one of a fine blue colour. I have not been unmindful of your hint about Scorpio, I am now *rummaging* the recesses of that constellation and find it full of beautiful globular clusters. A few evenings ago I lighted on a strange nebulæ, of which here is a figure! and since I am about it I shall add a figure of one of the resolvable nebulæ in the greater magellanic cloud. The equatorial is at last erected, and the revolving roof (upon a plan of my own) works perfectly well, but I am sorry to say that the nights in which it can be used to advantage are rare, even rarer than in England, as, in spite of the clearness of the sky, the stars are ill-defined and excessively tremulous. But a truce to astronomical details! though from time to time I shall continue to plague you

with them. . . . . Farewell; M. desires to add her
kindest regards to those of

Your affectionate nephew,

J. F. W. HERSCHEL.

The following letters from the Princesses of Hesse
and Dessau afford a pleasing memorial of the kind and
affectionate interest which they lost no opportunity of
expressing in Miss Herschel and her family.

HANOVER, *June* 10, 1834.

I yesterday received the enclosed note from my niece,
the Dowager Duchess of Anhalt Dessau, but felt too unwell
to send it as I could not write, which I wished to do, to
thank you also for your great kindness about the book.
My niece writes in *extasies* with your good nature. I am
glad to learn from our dear Sophy Beckedorff that you are
pretty well. I trust to be well enough soon to see you, but
I am still weak and unlike myself. It gave me very great
pleasure to learn that you have had fresh accounts of your
nephew, who, I pray God, may be prosperous in all his very
interesting and valuable undertakings.

I am happy of having this opportunity of assuring you of
the sincerity of my regard.

ELISE,

The Dowager Landgravine of Hesse,
born Princess of England.

TO MISS CAROLINE HERSCHEL.

[*Enclosure.*]

DESSAU, *June* 6, 1834.

Miss Caroline Herschel finds here the expressions
of my utmost gratitude for the great kindness to give me

the so very interesting work of her nephew, the worthy follower of a celebrated father.

The gentleman here, a Mr. Schwabe, to whom it was destined, looks with eager curiosity on the discoveries Mr. Herschel will make in the new regions of heaven he is now examining, and if she would be inclined, after receiving any interesting news, to make communication of it, it would always be accepted with the best thanks of

FREDRICA,
Duchess of Anhalt Dessau.

MISS CAROLINE HERSCHEL.

MISS HERSCHEL TO SIR J. F. W. HERSCHEL.

Sept. 11, 1834.

MY DEAREST NEPHEW,—

Your welcome letter of June 6 I received on the 19th August . . . . and I know not how to thank you sufficiently for the cheering account you give of the climate agreeing so well with you and all who are so dear to me, and that you find all about you so agreeable and comfortable, ... so that I have nothing left to wish for but a continuation of the same, and that I may only live to see the handwriting of your dear Caroline, though I have my doubts about lasting till then, for the thermometer standing 80° and 90° for upwards of two months, day and night, in my rooms (to which I am mostly confined) has made great havoc in my brittle constitution. I beg you will look to it that she learns to make her figures as you will find them in your father's MSS., such as he taught me to make. The daughter of a mathematician must write plain figures.

My little grand-nephew making alliance with your workmen shews that he is taking after his papa. I see you now in idea (memory ?) running about in petticoats among your father's carpenters, working with little tools of your own, and John Wiltshire (one of Pitt's men, whom you may perhaps remember), crying out, " Dang the boy, if he can't drive in a nail as well as I can!" but pray take care that he does not come to harm, and in your next tell me something of our little Isabella, too.

I thank you for the astronomical portion of your letter, and for your promise of future accounts of uncommon objects. It is not *clusters of stars* I want you to discover in the body of the Scorpion (or thereabout), for that does not answer my expectation, remembering having once heard your father, after a long awful silence, exclaim, " Hier ist wahrhaftig ein Loch im Himmel!"* and, as I said before, stopping afterwards at the same spot, but leaving it unsatisfied, &c. . . . .

About two months ago I was, for the last time, unfortunately, at the theatre, when Professor Schumacher and the Chevalier Kessel, of Danneburg, called on me. As soon as I came home I sent a note of invitation for the next levening, but had one returned informing me of their eaving Hanover next morning, and a promise of coming perhaps next summer. But I hear Struve is coming, and I hope I shall get a sight of him. The Emperor of Russia and the King of Denmark are cramming their observatories with astronomical instruments, &c., of all descriptions, made, I believe, some of them by Hohenbaum. . . . .

*   *   *   *   *

To my dear niece I beg you to give my best love and thanks for the kind arrangement to indemnify me for the loss of her dear letters, by charging her brother to inform me of all they know, &c., which, thank God, is hitherto of the most comforting nature.

With the most heartfelt wishes for the continuance of the health of you all,   I remain, &c., &c.,

C. HERSCHEL.

* Here, indeed, is a hole in heaven.

FROM J. F. W. HERSCHEL TO MISS HERSCHEL.

FELDHAUSEN, C. G. H., *Feb.* 22, 1835.

*   *   *   *   *

For my own part I never enjoyed such good health in England as I have done since I came here. The first coming on of the hot season affected me a little (odd enough with colds and rheumatisms), but it soon went off.

The stars continue to be propitious, and the nights which follow a shower, or a "black south-easter," are the most observing nights it is possible to imagine. I have swept well over Scorpio, and have many entries in my sweeping books of the kind you describe, viz., blank space in the heavens *without the smallest star.* For example :—

R.A. 16^h 15^m—N.P.D. 113° 56'—a field without the smallest star.
,,   16 19   ,,   116   3 — *Antares* (α Scorpii.)
,,   16 23   ,,   114 25 to 214° 5'—fields entirely void of stars.
,,   16 26   ,,   114 14   not a star 16^m—Nothing !
,,   16 27   ,,   114   0   ,,   as far as 114° 10'.

and so on.   Then come on the globular clusters, then more blank fields, then suddenly the Milky Way comes on as here described (from my sweep 474, July 29, 1834) :—

" 17^h 28^m, 114° 27'.—The Milky Way comes on in large milky nebulous irregular patches and banks, with few stars of visible magnitude, after a succession of black fields and extremely rare stars above 18th magnitude. I do not remember ever to have seen the Milky Way so decidedly nebulous, or, indeed, at all so, before."

Altogether the constitution of the Milky Way in its whole extent, from Scorpio to Argo Navis, is extremely curious and interesting. I have already collected a pretty large catalogue of southern nebulæ, for the most part hitherto unobserved, but my most remarkable object is a fine planetary nebula of a beautiful greenish-blue colour, a full and intense tint (not as when one says Lyra is a *bluish* star, &c.), but a positive and evident blue, between indigo-blue and verditer green. It is about 12″ in diameter, exactly round, or a *very* little elliptic, and quite as sharply defined as a planet. Its place is 11ʰ 42ᵐ R.A., and 146° 14′ N.P.D. My review for double stars goes on in moonlight nights, and among them I may mention γ Lupi and ε Chameleontis as among the closest and most interesting.

I have been hunting for Halley's comet by Rünker's Ephemeris in Taurus, but without success, though in the finest sky, quite dark, and with a newly-polished mirror. (By the way, I should mention that I have not had the least difficulty in my polishing work, and my mirrors are now more perfect than at any former time since I have used them.) My last comet hunt was Feb. 18. I shall, however, continue to look out for it. Pray mention this to Schumacher, who is Rünker's next door neighbour.

MISS HERSCHEL TO A. DE MORGAN, ESQ., SECRETARY OF
ROYAL ASTRONOMICAL SOCIETY.

March 9, 1835.

Sir,—

I return you many thanks for your communication of being chosen an Honorary Member of the Royal Astronomical Society, and beg you will do me the favour to convey my most heartfelt thanks to the honourable gentlemen of the Council for conferring so great an honour on me ; and only regret that at the feeble age of 85 I have no hope of making myself deserving of the great honour of seeing my name joined with that of the much distinguished Mrs. Somerville.

I beg you will believe me to remain, with great regard,
Sir, your most respectful and
obliged humble servant,
CAROLINE HERSCHEL.

HANOVER, *April* 2, 1835.

DEAR SIR,—

I feel very great gratification at recollecting that some twenty years ago I had the pleasure of being present when you were conversing at Slough with my dear brother, for it encourages me to address you now as an old friend, and I might almost say my only one, for death has not spared me one of those valuable men of the last century in whose society I had an opportunity of spending many happy hours, when they came to pass an astronomical night at Bath, Datchet, Clay Hall, and Slough. And I should now in the absence of my nephew (who would in my name have properly answered your kind letter for me) been much at a loss how to reply to yours of March 17. But I hope, dear Sir, you will have the goodness to return my sincere thanks to the Council of the Society for voting me a complete copy of their Memoirs. But, considering my advanced age and declining health, I think it best not to have them sent over to me, for it would cause me much uneasiness to leave them in the hands of those who could neither read nor understand them.

I suppose my nephew must have himself a complete copy of the Memoirs; but, if not, I beg you will give them to him, along with my love, as a keepsake from his affectionate and grateful aunt, the first opportunity you have to see him on his return.

Your kind information of the work with which you are at present engaged, touches a string which it has caused me no small trouble to silence; for whenever my thoughts return to those two or three years of which every moment that could be spared from other immediate astronomical business was, by my brother's desire, allotted for com-

paring each star of the British Catalogue with their observations in that *incorrect edition* of 1725, I feel always sorry
that want of time, and, perhaps, want of ability too, must
have been the cause of leaving many incorrections unnoticed. The work, however, was solely intended for the use
of my brother, who valued Flamsteed as an observer too
much to have made use of any other but the British Catalogue for determining the places of his newly-discovered
objects. N.B. We ought to remember that till the year
1790 and 1800, when Wollaston's and Bode's Catalogues
appeared, we had no other to go by, for those of Piazzi and
several other excellent observers were then not generally
known.

But, dear Sir, I ought to take leave of this to me interesting subject; for finding, about eight years since, that,
on account of the failure of my eyes and wretched health in
general, I should be unable to make further use of Flamsteed's works, I gave the three volumes, along with the
Atlas, Catalogue of Omitted Stars, &c., to the Observatory
of Göttingen, all marked throughout with what corrections
I knew of at that time; thinking they might be of use to
the observer there, and relieve me besides from the fear of
leaving them where they could not be appreciated, or an
attempt be made to comment on them, and perhaps have
made bad worse.

I wish (but almost fear life will not be spared me so long)
to see your new edition of the British Catalogue, therefore
beg you will favour me with a copy as early as possible. I
never knew that there was a Biography of Flamsteed's
existing, and trust you will favour me with the same as soon
as you can.

Any small parcels of astronomical papers will come to me
by favour of Herr Schumacher in Altona, who is so kind as
to send me his " Astronom. Nachrichten " regularly for

my amusement. And if you could send me the names of
the President and of the gentlemen of the present Council,
it would greatly oblige me.

I hope you will pardon my having intruded so long on
your time, but it has ever been my fault to be tedious in
expressing my thoughts on paper ; but will now only add
that, with great esteem and regard,

I remain, my dear Sir,

Your humble servant,

CAROLINE HERSCHEL.

FROM MRS. SOMERVILLE TO MISS HERSCHEL.

ROYAL HOSPITAL, CHELSEA,
April 16, 1835.

DEAR MADAM, —

I have sincere pleasure in availing myself of the
opportunity of writing to you which the Astronomical
Society of London has afforded me, by placing my name in
the number of Honorary Members, and greatly adding to
the value of that distinction by associating my name with
yours, to which I have looked up with so much admiration.

My object in writing is to request that you will accept
of a copy of my book on the Connexion of the Physical
Sciences, which is offered with great deference, having
been written for a very different class of readers.

I am proud of the friendship of your nephew, the worthy
son of such a father, who is succeeding so well in his
glorious undertaking at the Cape. I have seen a letter of
the 27th January, when they were all well and prospering.

I remain, dear Madam,

With sincere esteem,

Very truly yours,

MARY SOMERVILLE.

MISS HERSCHEL TO LADY HERSCHEL.

April 23, 1835.

My dearest Niece,—

   *   *   *   *   *

Your own dear letter arrived containing such a volume of joyful news, conveyed in the most kindest expressions, as if chosen for the purpose to cheer the heart of feeble age.

I was then not able (nor am I so now) to thank you as I could wish for your sparing so much of your valuable time and strength for the purpose of making me a partaker of your domestic happiness.

   *   *   *   *   *

I have now received in all five letters, two by your own hands, and three by my nephew. Each time after having read them over again they are put by, under thanksgiving to the Almighty, with a prayer for future protection.

. . . . Writing to my absent friends is one of the most laborious employments I could fly to when under bodily and, of course, mental sickness, for it is not impossible I might, instead of making inquiry about my little precious grand-nephew and the young *ladies* who play, sing, and sew so prettily, write, " O! my back. O! I have the cramp here, there, &c."

I had intended to keep a day-book to note down how and where I spent my time, and what was passing about me, which was to have served for yours and my nephew's amusement some day or other. But this I have given up long since, for seeing nothing but lapses of weeks and months where I could have given no better account of myself than that, after the fatigue of getting up and dressing, I fell asleep on the sofa, with a newspaper or other uninteresting subject in my hands, this would only have put me in mind of the useless life I am leading now.

But within the last two months I have been obliged to
exert myself once more to answer two letters, one to Mr.
De Morgan, the Secretary of the Royal Astronomical Society,
the other to Mr. Baily (who I suppose is President), for
they have been pleased to choose me, along with Mrs.
Somerville, to be a member (God knows what for) of their
Society. This, and receiving visits of congratulation (for
congratulate they must about all they find—what they call
promotion—in the zeitungen) has really somewhat disturbed
me, though Captain Müller and Mr. Hausmann I am
always glad to see ; with them I can talk about my nephew,
for they know him personally, and admire him. The winter
else has passed away rather heavily, because the Landgräfin
not being here, I had no other opportunity for seeing anything to put me in mind of England, but going to eight or
ten concerts, and those, ill or well, I never missed, for there
I was always sure to be noticed by the Duke of Cambridge
as his countrywoman (and that is what I want, I will be no
Hanoverian !), and then inquiries are made about my nephew
and his family ; even the little princess, twelve years old,
who sometimes when there, comes to give me her hand,
asking if I have had any letters from the Cape ; but now I
have seen the last of them, for the family go to England,
and will be absent for many months, and where may I be
when they return ? But Sunday night I sat a full hour on
the sofa with the Duke at Mrs. Beckedorff's, where I go
Sundays from seven to nine, where there is nobody but the
female part of Mrs. B.'s family, and another old lady, who
was absent on account of being not well. Of this our meeting the good Duke knew all along, and good-naturedly came
to join our gossip.

Here I have filled my paper with talking of nothing but myself, because I know that my nephew corresponds with *all* scientific men in Europe, for I hear frequently of extracts having

appeared in the papers (of his communications) by Struve, Littrow, &c., and should suppose he will also know what is done at *our Society, of which I now am a fellow!* and is of course acquainted with what Mr. Baily mentioned in his letter to me, that at the public expense a new edition of Flamsteed's work is now in print, and that papers have been found at the Royal Society containing a biography by Flamsteed's own hands, which—but here I transcribe what Mr. B. writes:—"I lament very much, in common with every friend of science, that Newton's name is mixed up with transactions that show him in a different light from that in which we have generally received his character. But justice to Flamsteed's memory would not allow me to suppress any portion of his autobiography."

Now we talk of biographies, I have no less than nine of my poor brother, and heard of two more, one by Zach, which I shall try to get sight of. There is but one or two which are bordering on truth, the rest being stuff, not worth while to fret about. The best is accompanied with a miniature of Reberg's *bad* copy; but I have ordered a lithograph copy to be taken from the portrait by Artaud; if it turns out correct I will send two copies as soon as they come out.

God bless you both, and the dear children, my best niece.

Ever your most affectionate aunt,

CAR. HERSCHEL.

MISS HERSCHEL TO P. STEWART, ESQ.*

May 25, 1835.

* * * * *

Let the time come whenever it may please God, I leave

---

* A brother of Lady Herschel's. This gentleman and his brothers were in the habit of writing to Miss Herschel during her nephew's absence at the Cape, keeping her informed of the latest news, and showing her every kind and thoughtful attention.

cash enough behind to clear me from *all* and *any* obligations to all who *here* do know me. Even the expenses of a respectable funeral lie ready to enable my friend Mrs. Beckedorff, and one of my nieces (the widow of Amptmann Knipping,* who lately came to settle at Hanover) to fulfil my directions.

I hope you will pardon my troubling you with such doleful subjects, but I wish to show you that my income is by one third more than I have the power to spend, for by a twelve years' trial I find that I cannot get rid of more than 600 thl. = £100 per year, without making myself ridiculous.

MISS HERSCHEL TO LADY HERSCHEL.

HANOVER, *August* 6, 1835.

MY DEAREST NIECE,—

I dare not wait any longer for a return of better spirits, *such* as in which I should like to reply to my nephew's dated February 22nd, and yours of May 19th, for I fear if I do not at least acknowledge the receipt of them, I shall not be gladdened again by such delightful descriptions of your health and healthful situation, and my nephew's contentment with the successful progress he is making in his intended observations.

At first, on reading them, I could turn wild, but this is only a flash, for soon I fall in a reverie of what my dear nephew's father would have felt if such letters could have been directed to him, and cannot suppress my wish that *his* life instead of *mine* had been spared until this present moment; for what immense and wonderful discoveries have

---

* This lady, the daughter of Dietrich Herschel, proved a most true, affectionate, and trustworthy friend to the last. See her letter on Miss Herschel's death.

not been made within these thirteen years, chiefly by his own son, or son's suggestion!

But I must stop here and turn to more earthly and indifferent subjects (though they ought not to be called indifferent neither), for in the first place I have to return my thanks for no less than three dozen of Constantia wine, but this I shall do but with a very bad grace, for ever since the 11th of May, when I received my nephew's letter, I have been in the fidgets about the trouble he and his friends must have had before such a thing could reach me. . . . . I feel more reconciled after unburdening myself of some of this weighty concern by making presents to all who love and esteem you so truly, and after setting apart a portion, according to Captain Müller's advice, with which you may be treated when at your return you may perhaps visit Hanover again, there remains more than ever I can get through with, for I am very desirous to spin out the thread of my life till you return home. And I know it is a mistaken notion that old folks want more of what they call comfort than young ones. It is not very easy to find out what will convey comfort in general. . . . . I, for instance, know of no other comforts like those I derive from yours and my dear niece's letters. Her last leaves me nothing to wish for. . . . .

*   *   *   *   *

You compliment me on having a steady hand, but if you were to see the blotting I make before I can make it hang together (when I am *composing*, as it were, a letter) you would not say so, and, after all, it will cause you some trouble to understand me, for the letter begins to my dear niece, and soon after I find myself talking to you. . . . .

FELLHAUSEN, *Oct.* 24, 1835.

DEAR AUNT,—

The last accounts we have of you are that you are
elected a member of the Astronomical Society, and that to
keep you in countenance, and prevent your being the only
lady among so many gentlemen, you have for a colleague
and sister member, Mrs. Somerville. Now this is well
imagined, and we were not a little pleased to hear it. May
you long enjoy your well-earned laurels !

As I presume our news will interest you more than comments upon what goes on in Europe, in the first place be it
known to you, that we are all well and, thank Heaven, happy.
The children, one and all, thrive uncommonly. . . . . The
stars go on very well, though for the last two months the
weather has been chiefly cloudy, which has hitherto prevented me seeing Halley's comet. Encke's (*yours*) escaped
me, owing to trees and the Table Mountain, though I cut
away a good gap in our principal oak avenue to get at it.
However, Maclear, at the Observatory, succeeded in getting
three views of it with the fourteen-foot Newtonian of my
father's (the Glasgow telescope) on the 14th, 19th, and
(?) 24th of September. If you have an opportunity of
letting this become known to Encke, pray do so—(I shall
write to him shortly myself). It was *in* or *near* the calculated place, but no measures could be got.

I have now very nearly gone over the whole southern
heavens, and over much of it often. So that after another
season of reviewing, verifying, and making up accounts (re-
ducing and bringing in order the observations) we shall be
looking homewards. In short, I have, to use a homely
phrase, broken the neck of the work, and my main object
now is to *secure* and perfect what is done, and get all ready

to begin printing the moment we arrive in England; or, if that is not possible, at least to have no more calculation to do. . . . .

FROM H.R.H. THE DUKE OF CAMBRIDGE.

HANOVER, *Nov.* 19, 1835.

The Duke of Cambridge hastens to acknowledge the receipt of Miss Herschel's very obliging note, and to return his many thanks for her attention in sending him some of the Constantia she has lately received from her nephew. He seizes this opportunity of assuring her of the satisfaction he felt at hearing that Mr. Herschel and his family were in good health, and he sincerely hopes that the climate of the Cape will agree with them.

FRANCIS BAILY, TO MISS HERSCHEL.

37, TAVISTOCK PLACE, LONDON,
Jan. 29, 1836.

MY DEAR MADAM,—

I forwarded some time since, to Professor Schumacher, a copy of my "Account of Flamsteed," to be sent to you; and which he says was duly transmitted. I am anxious to know whether it has arrived safe, for, as only a limited number of copies were printed (which are *all* distributed) it cannot be purchased.

I have been the more desirous that *you* should have a copy, because there is no one that has taken so much pains to elucidate and explain the works of Flamsteed as yourself, and therefore I am bound in gratitude to see that you are put in possession of a copy of the work.

I shall take this opportunity of stating that I hear occasionally from your nephew at the Cape of Good Hope, and that the last accounts confirmed his continuance in good health, and his enjoyment of the pleasures of the fine climate in which he is placed.

I remain, my dear madam, with the assurance of my best respects, and my best wishes for your health and happiness,

Your very obedient servant,

FRANCIS BAILY.

MISS HERSCHEL TO F. BAILY, ESQ.

HANOVER, *Feb.* 15, 1836.

DEAR SIR,—

I am quite at a loss for terms in which to apologize for having neglected to acknowledge the receiving of your valuable Catalogue and biography of our dear ill-used Flamsteed, which was forwarded to me by the usual kindness and punctuality of Prof. Schumacher on the 9th October last. The same packet also contained Mrs. Somerville's second edition "On the Connexion," &c., accompanied by a kind note, dated as far back as April 16th, which, to my sorrow, is also still left unanswered on account of illness, and in the hope that when the days are somewhat longer (my eyes fail me), and that with the return of spring I might perhaps regain some small portion of strength—but I doubt.

The parcel also contained duplicates of my nephew's second series, and on the satellites of Uranus, and I must trust that on his return he will convey my grateful thanks to you, sir, and the gentlemen, for all the kind attention conferred on me during his absence. My last letter from the Cape is dated October 24th, and I am much gratified by your kindness in having informed my nephew of the wish I have that the volumes of the Royal Astronomical Society's publications voted to me might be kept for him, and he seems much pleased with the arrangement. I therefore would recommend them to your obliging care till his return.

The volume of your " Account of Flamsteed " must be my
companion to the last, but I will take care it shall be safely
delivered to my nephew.

If I will not lose another post I must conclude with the
assurance of ever remaining with great regard,

My dear Sir,

Your much obliged and humble servant,

C. HERSCHEL.

SIR JOHN HERSCHEL TO MISS HERSCHEL.

March 8, 1836.

DEAR AUNT,—

Maggie desires me to finish this for her, but she has
not left me room to write at length. So I will only devote
this space to one point in your last letter which requires reply.
I have not got Gauss's apparatus, and I am not sufficiently
acquainted with his method of observing to construct one for
myself.   Besides which it is quite out of my power to undertake any extensive series of observations, being anxious to
get home, and having still so much to do, both in observation and reduction, that I really shall hardly be able to accomplish all I have already in hand.   This comet [Halley's]
has been a great interruption to my sweeps, and I *hope* and
fear it may yet be visible another month.   Unluckily when
I sailed from England I left all my volumes of Poggendorff
and the " Nachrichten " behind me, and none of the former
and very few of the latter have reached me here.   I fear it
is now too late to send home for anything, and I have two
series of observations, viz., of the comparative brightness
of the southern stars, and of the photometric estimation of
their magnitudes—the former just commencing, the latter
not yet begun, which I *must* do.   Pray explain this to
Gauss. . . . . Astronomical news I have little, but one

thing very remarkable I must tell you, γ Virginis is now *a single star* in both the twenty-foot and the seven-foot equatorial ! ! !

Your affectionate nephew,

J. F. W. HERSCHEL.

MISS HERSCHEL TO SIR J. F. W. HERSCHEL.

HANOVER, *June* 29, 1836.

MY DEAREST NEPHEW,—

I do not know where to begin, for I see it is nearly a twelvemonth since I gave some account of myself, and in all that time never returned my thanks for the three letters I received. . . . . I have a great deal to say, and will begin with accounting for my long silence, by confessing that I have throughout the whole winter been too ill to do anything besides nursing myself, and putting myself in a condition to appear before strangers, which I am not able to do till after twelve or one at noon, and the time which I wanted to rest after my exertion and getting my breakfast was generally taken up by pacifying the *gulls* about the foolish paragraphs they had been reading the night before in the Clubs. I never read, or would read, any of them, but when I heard of anything appearing rational concerning you, I copied or procured the paper for myself, and then I found among the rest a letter of yours to Professor Plana, in Turin, dated December 28th, 1834. And not being able to do anything of use to you myself, I begged Capt. Müller to cause those observations of June 21st, &c., to be made by somebody here in Hanover, and the enclosed letter will, I hope, meet with a gracious reception. . . . . I believe Dr. Heere will not fail the next equinox to be at his post, and you may hear more of him.

Capt. Müller is at present with Gauss, and will deliver all your messages personally, for you must know I beware of corresponding with all those *known ones* if I can possibly help it, and have through his hands sent copies of your father's likeness to Struve, Schumacher, Gauss, Bessel, Encke, Olbers, &c. Gauss sent me word it was hung up in his library. Encke sent me a very pretty letter of thanks. . . . . That sending is an ugly thing. Mrs. Somerville sent me her book with a letter dated April 16th, which I received October 9th, coming along with Mr. Baily's publication, *presented by the Lords Commissioners of the Admiralty to Miss Herschel.* You cannot think how agitated I feel on such occasions, coming to *me* with such things !— an old poor sick creature in her dotage. . . . . I was going to say something yet of Mr. Baily's labours, but the paper is at an end ; but I hope you will now soon read in your own library at Slough what the " Quarterly Review," No. CIX., says, and what your Cambridge friend Whewell and others have said—in short, Newton remains Newton ! God bless my dear nephew and niece ! . . . . My heart is too full—I can say no more than that

I am your affectionate aunt,

CAR. HERSCHEL.

MISS HERSCHEL TO LADY HERSCHEL.

HANOVER, *October* 20, 1836.

MY DEAREST NIECE,—

From June 14 to October the 1st, and not *any the least account*, was rather too much for me to bear, especially during the months when those few friends who sometimes cheer me by a friendly call had all left the town to make summer excursions. . . . .

I have a few memorandums for my nephew, and will for the present take leave of my dear niece with my most heartfelt wishes that every future account with which I may yet be blessed from her dear hand may be like the last.

*   *   *   *   *

. . . . I have four complete years of the "Astronom. Nachrichten" ready bound for you. . . . . I wished to give you the number of the paper (but cannot find it again) where Bessel speaks of Saturn's satellites, but my eyes are so dim, and I am too unwell for doing anything. I will therefore only say he has seen the 6th but not the 7th, the ring being in the way. In No. 293, two of Bessel's assistants, Beer and Mädler, say a great deal about the observations of your father, but that goes all for nothing. I will only say in general that he did in one season more than any one else could have done, and would have resumed the *hunt* the next fifteen years if nothing had interfered. And the Georgium Sidus was followed as long as anything could be obtained from that planet, and it will yet be some twenty years before he will be in that favourable situation in the ecliptic where he was at the time when the satellites were discovered.

I have seen Struve's Catalogue of Double Stars, wherein I find he agrees with your and your father's observations. . . . . Do not think, my dear nephew, that I would expose myself so as to say a word about these things to anybody else, but to you I cannot help letting it out when I am nettled.

I must leave off gossiping, else I shall not get this letter away, in which you will find Dr. H.'s barometrical observations, which I received a few days ago. . . . .

SIR J. F. W. HERSCHEL TO MISS HERSCHEL.

CAPE OF GOOD HOPE, *Jan.* 10, 1837.

MY DEAR AUNT,—

*   *   *   *   *

I am now at work on the spots in the sun, and the general subject of solar radiation, which you know occu-

pied a large portion of my father's attention. The present
is an admirable opportunity for studying these things, as
the sun is infested now with spots to a greater degree than
ever I knew it, and they are arranged over its surface in a
manner singularly interesting and instructive. The sky
here is so pure and clear in our summer that it would be
a shame to neglect such an opportunity of making experiments on heat, and accordingly I have been occupied in
the December solstice in determining the constant of solar
radiation, that is to say, the absolute quantity of heat sent
down to the earth's surface from the sun at noon, or at a
vertical incidence.

I do not think I have ever mentioned to you a remarkable and splendid instance of liberality on the part of His
Grace of Northumberland, who has taken upon himself to
defray the expenses of publishing my observations at the
Cape, and that in a manner the most delicate and considerate imaginable. In consequence " my book " will appear,
when it does appear, under his auspices, and I hope it will
not do discredit to his munificence. This is not the only,
nor the most remarkable, instance however, of his attach-
ment to the cause of science, and his disposition to pro-
mote and support it.

MISS HERSCHEL TO SIR J. F. W. HERSCHEL.

March 30, 1837.

*   *   *   *   *

. . . . I have for the last five months been in continued fear of losing Mrs. Beckedorff (to whom I could
confide all my grievances). She is worn out with a cough
and breaking up of constitution, and we but seldom can
come together, which is when I am able to cross the street
to go to her. . . . . I experience a daily increase of pain
and feebleness, so that I am (at least during this severe

weather) totally confined to my solitary home; and what is
worse, my eyes will not serve me to amuse myself with
reading. But what business had I to live so very long?

FROM SIR J. F. W. HERSCHEL TO MISS HERSCHEL.

FELDHAUSEN, *May* 7, 1837.

. . . . I will try to entertain you with some celestial
affairs in which it is delightful to find you still taking so
much interest. As you allude to Saturn's satellites in your
letter of October 20, I must tell you that I have *at last* got
decisive observations of the sixth satellite (the farthest of
my father's new ones). I had all but given the search up
in despair, when no longer ago than *last Thursday* (May 4th
inst.), being occupied in taking measures of the angles of
position of the five old satellites with the twenty-foot and a
polished new mirror, behold, there stood Mr. Sixth! a little

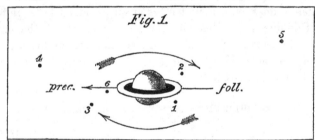

Fig. 1.

short of its preceding elongation. I have kept it well in
sight from 14^h 26^m Sid. T. till 16^h 35^m, in which time it had
advanced visibly in its orbit from *bdoro*, the line of the
Ansæ (as in figure) *to above.* In

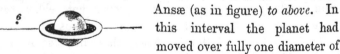

this interval the planet had
moved over fully one diameter of
the body towards the preceding side, and, therefore, had

it been a star, must have passed over it, whereas it preserved the same apparent distance all the while from the edge of the ring. (N.B. Saturn not very far from the zenith on merid.)

Next night, *Friday, May 5,* Saturn most gloriously seen : quite as sharp as any copper-plate engraving, with power 240 and full aperture. All the five old satellites seen and measured, being now on the opposite side. Now considerably short of its greatest *following* elongation ; distance just

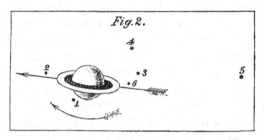

as before; and, as on Thursday, it was kept in view long enough for Saturn to have left it behind by its own motion had it been a star. The change of situation agrees perfectly with the period $1^d\ 9^h$, which is also the reason why it was not seen May 5th, being on that night near its inferior conjunction. So this is *at last* a thing made out. As for No. Seven I have no hope of ever seeing it.

If your eyesight will not suffer from it, do write to Bessel. I am sure he will be interested by this observation, as he is the only astronomer who troubles himself about the system of Saturn. I shall myself write to him shortly about it, but should like to have a few more observations.

So now farewell once more, and, with many kind remembrances to all Hanoverian friends,

Believe me, your affectionate nephew,

J. HERSCHEL.

U

MISS HERSCHEL TO SIR JOHN HERSCHEL.

HANOVER, *June* 11, 1837.

*   *   *   *   *

. . . . From Mr. Schumacher I receive each paper as it comes from the press, but always with a feeling of uneasiness, because I am not one of those who can contribute anything to their valuable communications, nor even understand all which my defective eyes allow me to read. But they interest me exceedingly when I think what you will say. For instance, to a paper of twenty-two quarto pages, by Bessel, " Über den Einfluss der Unregelmassigkeiten der Erde, auf geodetische Arbeiten und ihre Vergleichung mit den astronomischen Bestimmungen." * Perhaps you may have received these papers before this reaches you, but if any are lost by the way, I collect them for you ; but I fear I shall not see the day of all the wonders coming to light when *you* return with your budget. . . . .

. . . . I must conclude, for writing at any time makes me sad ; and since I began this letter the notice of the death of our King has arrived, and the Duke of Cumberland has been this day proclaimed King of Hanover. It makes me feel as if I was doubly separated from England, for your King is now no longer my King. And we lose the Duke of Cambridge, who was ever so kind to me wherever he saw me. Last winter he introduced me to his brother, then Duke of Cumberland, who was here on a visit, at the concert, who spoke to me of you first as my son, but recollected himself that I was only aunt. . . . .

*   *   *   *   *

I had illuminated my front rooms with twenty candles (snuffed them all myself, for Betty was out to see the show) on the evening of the King's arrival, and so I shall again

* " On the Influence of the Irregularities of the Earth on Geodetic Operations, and their Comparison with Astronomical Determinations."

next Saturday or Sunday, when the Queen is expected.
More I cannot do ! . . . .

. . . . My head becomes crowded with melancholy fore-
bodings of my not lasting so long as to hear of your safe
return to your home and the friends which I think are only
to be found in happy England; so, instead of tracing my
gloomy imaginations on paper, I go to sleep till Betty rouses
me with a cup of coffee. . . . . But all I hear of you is
told in a tone of admiration, &c., &c., and it is felt by me
like a drop of oil supplying my expiring lamp.

J. F. W. HERSCHEL TO MISS HERSCHEL.

CAPE OF GOOD HOPE,
Sept. 7, 1837.

MY DEAR AUNT,—

*   *   *   *   *

I need hardly say how much we are rejoiced to see
your handwriting once more, though that joy is damped
by your complaints of winter indisposition. And such a
winter! by all accounts. May this prove a better! and
may we hope to find you in no worse health and spirits when
we come to see you *next summer* in Hanover. For so, if it
please God to lead us safe home, according to our present
altered plans, we most assuredly propose to do.

I say our *altered* plans, for you know our intention was
to have embarked next March for Rio Janeiro, and there to
have spent two or three months, after which to have taken
passage in the Brazilian packet for England, which would
have probably detained us till October, and have rendered a
visit to Hanover that season impracticable. But by striking
off this Brazilian trip, and taking our course directly homewards, so much time will be saved, and all the rest of our
domestic arrangements become so much simplified that it
seems like finding a treasure, as a fund of time will thereby
be placed at our disposal, the first fruits of which, as in

all love and duty bound, we have determined to devote to you; or rather, I should say, that, when in talking over with Margaret all the *pro's* and *con's* of the question, whether to return home direct, or *viâ* Brazil?—*this* consideration at once decided it in favour of the direct course, her desire to see you outweighing every consideration of amusement or temporary gratification which a visit to Rio could offer. So now be sure, dear aunty, and keep yourself well, and let us find you in your best looks and spirits; and, although what you say respecting our good Mrs. Beckedorff's health is somewhat deplorable, yet I will indulge the hope that she too will perform a part in the *dramatis personæ* of that happy meeting. Meanwhile, as the time of our departure hence approaches, we shall take care and apprise you of all our movements, respecting which it is impossible at present to speak more precisely.

FROM H.R.H. THE DUKE OF CAMBRIDGE.

CAMBRIDGE HOUSE, *May* 18, 1838.

MY DEAR MADAM,—

Having just been informed by the newspapers that your nephew is safely landed in this country, I hasten to write you a few lines by this night's mail to congratulate you most sincerely on this event, which I know will give you pleasure.

I am unable to send you any further details about him or his family, as I am not aware if he is arrived as yet in town, and should this not be the case, my letter will perhaps be the first to give you this welcome news, which I shall certainly be delighted at.

I trust you continue enjoying your health; and with best wishes, &c., &c.,

Yours most sincerely,

ADOLPHUS.

SIR JOHN HERSCHEL TO MISS HERSCHEL.

LONDON, *May* 20, 1838.

Here we are, my dear aunt, at last, safely landed and housed, all in good health and, as you may suppose, in good spirits at our return. We ourselves and our six little ones were very comfortable during our nine weeks' voyage in the good ship *Windsor*, which is lying snug and sound in the river at Blackwall, with all our things on board, telescopes and all (as well as the astronomical results of our expedition). We left our ship, however, at the entrance of the Channel, and got to London in a steamer under the flag of King Leopold, of Belgium, which, having been to Glasgow to take in her machinery, was returning without passengers, not yet being fitted up for their reception. This was a most opportune and unexpected piece of good fortune, as I assure you we found most sensibly, by the non-arrival of the ship till this morning, having been four days longer at sea, beating about against contrary winds. I have more particulars to tell than would fill this paper, which I must reserve till our meeting, which will not now be longer delayed than is indispensable for getting our baggage on shore, and passing it through the Custom House, and transporting it by a barge to Windsor, and so to Slough. I hope and trust to find you as well in health as your two letters to John and Mary Baldwin allow us to suppose. . . . . .

The visit promised in the foregoing letter was paid in July, when Sir John Herschel, accompanied by his little son, spent a few days with his aunt, whose intense anxiety as to the proper treatment of her little grand-nephew—his sleep, his food, his playthings—kept her in a constant state of alarm on his account. "I," she writes, "rather suffered him to hunger than would let him eat anything hurtful; indeed, I would not let him eat anything at all without his papa was present." Great as was the joy of the dear venerable lady to rest her aged eyes once more on almost the only living being upon whom she poured some of that wealth of affection with which her heart never ceased to overflow, it is on the disappointments and shortcomings of those few precious days that she dwells; and, if she could have felt resentment towards her nephew, it would have been roused by the abrupt termination of his visit. Her lamentations are piteous. Solely with the intention of sparing her feelings, her nephew went away without letting her know the exact time beforehand of his departure, and made no formal leave-taking, when he bade her good-night to return to his inn. To her infinite dismay and distress, she found that he and his son had quitted Hanover at four o'clock on the following morning. It was kindly intended, but it was a mistake that gave intense pain. Her introduction to her little grand-nephew is described as follows by his father, H. Herschel :—

. . . . "Now let me tell you how things fell out. Dr. Groskopff took Willie with him to aunty, but without saying who he was. Says she, 'What little boy is that?' Says he, 'The son of a friend of mine. Ask him his name.' However, Willie would not tell his name. 'Where do you come from, little fellow?' 'From the Cape of Good Hope,'

says Willie. ' What is that he says?' 'He says he comes
from the Cape of Good Hope.' 'Ay? and who is he?
What is his name?' 'His name is Herschel.' 'Yes,' says
Willie, 'William James Herschel.' 'Ach, mein Gott! das
ist nicht möglich; ist dieser meines Neffen's Sohn?' And so
it all came out, and when I came to her all was understood,
and we sat down and talked as quietly as if we had parted
but yesterday. . . . .

" Groskopff, by the way, was recounting a strange feat
which, to give you some notion of the *sort of person* (*par
rapport au physique*), she performed, not longer than half a
year ago. Remember it is a person of eighty-eight or
eighty-nine of whom we are speaking. Well! what do you
say of such a person being able to put her foot behind her
back and scratch her ear, in imitation of a dog, with it, in
one of her merry moods?"

The "Day-Book," which, as already stated, had been
recommenced in the year 1833. The first volume of
the new Day-Book concludes in May, 1837, with
comments on Baily's account of Flamsteed, and recollections of days spent at Greenwich in 1799, when she
had seen and wondered at the piles of manuscripts
accumulated there. "Dr. Maskelyne was not indifferent to the stores of observations of his predecessor,
for he even attempted to make *me* undertake the examination of some of Halley's scribblings on fragments
of waste paper [to see if they] might not belong to
some star or other. But such things cannot be done
in a moment, and the parcel was restored to its dusty
shelf. Poor Dr. Maskelyne had but one assistant,

with a salary of £70 a-year, whom I once heard lament
that all the planets happened to pass the meridian in
the night-time!"

The entries are chiefly of the numerous visitors she
received, but there are frequent intervals of several
months when illness or disinclination to write prevented her continuing her Journal regularly. The
English Quarterly and Monthly Reviews and newspapers, and James's novels, supplied her with constant
reading, and every allusion to her brother's or her
nephew's labours is carefully noted. It is evident that
she still was in the habit of taking ample notes of any
book that interested her, in spite of complaints of the
growing failure of sight, and that, when tolerably well,
no day was considered altogether satisfactory which
was passed in solitude. It was in May, 1833, that
she moved to No. 376, Braunschweiger Strasse, and
here she continued to dwell for the remainder of her
days.

MISS HERSCHEL TO LADY HERSCHEL.

HANOVER, *July* 30, 1838.

MY DEAREST NIECE,—

I hope that when you receive this my dear nephew, with
his precious charge (little William), will be safely restored
to your longing arms, and that he may have found you, with
all the little family, in perfect health. I wish to be assured
by a few lines from your dear hands as soon as possible, for
I cannot divest myself of a fear that the botheration and
intrusion of some of the stupid Hanoverians must have

been very inconvenient to him. To which may be added the change of weather from excessive heat to very cold and wet, to which at this present moment (as far as I know) they are still exposed, for I think they must be now in Hamburg. . . . .

SIR J. F. W. HERSCHEL TO MISS HERSCHEL.

LONDON, *Aug.* 6, 1838.

My dear Aunt,—

Willie and I arrived in London safe and hearty on Friday night about eight o'clock, and I am happy to say we found all here quite well—both mamma and all the little folks, who, as you may easily imagine, were in great joy, and full of enquiries about you and about all our adventures in foreign parts. Grandmamma Stewart, and all her circle also, with exception of poor James S. (who is, however, much better, and we hope permanently), are well, and join us in kind enquiries after you. I found here my cousin, Thomas Baldwin, and his excellent and most amiable wife. Cousin Mary had left us, and was returned to Anstey.

I found Dr. Olbers well, and have to thank you, in his name, for the Cape wine, a bottle of which was produced at dinner the day I dined there. I assure you it was drank in good company, being associated (*not mixed*) with Hock of 240 years of age!! Dr. O. is weak and corpulent, but is otherwise in the full enjoyment of his mental faculties, and in good spirits.

I could not persuade myself to encounter a regular parting with you, and, in fact, I found the distance to Bremen so much greater, on enquiry, than I had fancied it, that it was necessary to leave Hanover at four a.m., which, of course, prevented all further meeting. We shall be most anxious to hear from you. M. will write in a day or two (and so will the children) to thank you for all your kind remembrances of them, and for the many pretty and valuable things you have sent; and till then, believe me,

My dear aunt,

Ever your affectionate nephew,

J. F. W. HERSCHEL.

MISS HERSCHEL TO SIR J. F. W. HERSCHEL.

HANOVER, *Aug.* 21, 1838.

MY DEAREST NEPHEW,—

By the arrival of your letter of the 6th I was relieved from my fears for the safety of you and your dear little fellow-traveller, almost a week sooner than I had reason to hope.

*   *   *   *   *

. . . . I had so longed for a few hours of confidential conversation with you which would have spared me the unpleasant task of writing about earthly matters. . . . . My good neighbours came to wish me joy, and congratulate me on having seen my glorious nephew and his son (who has left no few admirers behind, I can tell you).

Dr. Mühry has lost a sister, a solitary old maid, like myself, whom they could not leave till she was buried. But she was in some respects better off than I, for I found it necessary to order all these matters myself. Miss Beckedorff and Mde. Knipping will at my death have to deliver a sealed packet to Dr. Groskopff, my executor, in which, on his opening in their presence, he will find the means requisite for discharging all the items specified in an enclosed memorandum of directions. Such matters I had wished to talk over with you, thinking it not unnecessary you should know a little about the way in which I have always managed my affairs. As soon as I was left to myself, in the year 1788, I kept a book strictly accounting for my expenses, which was to serve as a voucher of the orderly life I led.

But being frequently under the necessity of assisting one or other of my, as I thought, poor (but say extravagant) relations, I began to keep a spare box, by way of showing to what extent I have thus robbed myself. . . . . I am sorry to trouble you with such details, but I find myself so unwell at present that I cannot rest till I have cautioned you not to ask any question about me of any one, for nobody knows anything about me—my confidence in Mrs. Beckedorff, even, can only be partial, as we can only see each other so seldom.

*   *   *   *   *

MISS HERSCHEL TO LADY HERSCHEL.

HANOVER, *Sept.* 24, 1838.

*   *   *   *   *

I see by the postscripts you directed my nephew to add to your letter that you know exactly what will make his poor old aunt happy ; and I must beg you to make my peace with my dear little William, for I fear the angry looks I gave him when seeing him climbing too high on an open window two stories above the pavement, can have left no favourable impression on his recollection. Unfortunately we could not converse together : he talked too soft and quick for me (I do not hear so well as formerly), and my mixture of German and English was not intelligible to him. . . . . Had the knitting with beads been known forty years sooner, it would have been one of the accomplishments with which I came, at the age of twenty-two, into England in 1772, for there was no kind of ornamental needlework, knotting, plaiting hair, stringing beads and bugles, &c., of which I did not make samples by way of mastering the art. But as it was my lot to be the Cinderella of the family (being the only girl) I could never find time for improving myself in many things I knew, and which, after all, proved of no use to me afterwards, except what little I knew of music, being just able to play the second violin of an overture or easy quartette, which my father took a pleasure in teaching me. N.B. When my mother was not at home. Amen. I must think no more of those times, only just say I came to Bath with a mind eager to learn and to work, and never changed my mind till I came here again, but now I can no more. . . . . One thing I must tell my nephew, which is, that I hope I have found a deserving protector of my sweeper in Director Hausmann, and I hope either himself or his son will find us a few comets with it yet. He is a constant visitor of mine.

SIR J. F. W. HERSCHEL TO MISS HERSCHEL.

SLOUGH, *Nov.* 26, 1838.

My DEAR AUNT,—

I have received a letter from Sir Wm. Hamilton, Astronomer Royal, Dublin, informing me that the Royal Irish Academy have elected you an honorary member of that body. The diploma is by this time on its way to my care, and I will, so soon as I receive it, take the very first secure opportunity of transmitting it to you.

Yesterday I received your most welcome letter and Mr. Bagulawski's in one. I wrote to him some time ago relative to Halley's comet. He seems a very diligent observer, and I am glad you have seen him.

Your letter of September 24th, with its numerous dates, was like a little diary, and almost made us fancy ourselves with you in Hanover. . . . .

I am sorry to see, on looking at my banker's account, that you have *not* (as you promised to do) drawn on Cohen for the £50 of this half year. Pray do, and that soon, or I shall be sadly disappointed.

We have got a most excellent president for the Royal Society in the Marquis of Northampton. He presided at the anniversary dinner on the 30th, and did the honours with great credit.

A Copley Medal was awarded to Gauss for his researches, theoretical and practical, on the subject of terrestrial magnetism.

MISS HERSCHEL TO SIR J. F. W. HERSCHEL

HANOVER, *Dec.* 17, 1838.

MY DEAR NEPHEW,—

First and foremost let me dispatch what may be called business. In the first place, I thank you for your kind letter and communication of having so great an honour conferred on me as to be admitted an honorary member of the Royal Irish Academy. I cannot help crying out aloud to myself, every now and then, *What is* THAT *for?* Next I must beg you to return my thanks in what words you think proper I should express them, and if you will only send me a copy of the diploma, and keep the original along with my other trophies, allowing them perhaps a corner in some such box as that your dear mother had for suchlike things, for I have no other desire but to be remembered by you and Lady H., and your children, for yet awhile. . . . .

. . . . It is a long while since you asked me if I wanted any of my Indexes to Flamsteed's Catalogue of omitted stars. If there should yet be any left, I could wish to have one or two; for you hinted to me I might leave Baily's work to the "Archives" here, which I intend to do, and then I should like to give an Index along with it.

*   *   *   *   *

MISS HERSCHEL TO SIR JOHN HERSCHEL.

HANOVER, *Jan.* 7, 1839.

I see, to my sorrow, that my letter was not come to hand
at the time when you directed the parcel with the diploma,
which was sent me on the 2nd of January, accompanied by a
note from the President, which I beg you will answer for
me, and for that purpose transcribe here the same :—

OBSERVATORY, DUBLIN,
Dec. 4, 1838.

" MADAM,—

" In transmitting to you the accompanying Diploma
from the Royal Irish Academy, I wish to be allowed to add,
as I thus do, the expression of my own high sense of your
services to Astronomy, and of the eminent degree in which
you have deserved the present testimonial.

" I have the honour to be, Madam, &c., &c.,

" WILLIAM RANAN HAMILTON,

" P.R.I.A."

MISS HERSCHEL TO SIR JOHN HERSCHEL.

HANOVER, *Dec.* 1, 1839.

DEAR NEPHEW,—

Do not you think I have been very good to let the
most dismal month in the year pass without troubling you
for accounts of the progress my dear niece is making in her
recovery ?

My dear niece said once, I should write often, and in few
lines inform her how I go on, so I must say—I get up as
usual every day, change my clothing, eat, drink, and go to
sleep again on the sofa, except I am roused by visitors;
then I talk till I can no more—nineteen to the dozen!
N.B. I don't tell *fibs*, though they may not always like what
I say.

I have been twice at the concert, and each time been

honoured with a wie gehts?* by His Majesty, and the notice of many acquaintances whom I have no opportunity of seeing elsewhere, the public concerts being the only place where I can go with the least trouble to myself or others. You say when I talk of the gelehrten then all goes well, but I know nothing about them. . . . .

But one piece of news I must tell you, which is, that a fortnight after Dr. Mädler had been the conductor of Mde. Witte (the Moon) and her daughter to the meeting at Pyrmont, I received two cards, the one, " Professor Dr. Mädler," under it, "Minna Witte-*Verlobt*.†" The reason Madame Witte gives for this hasty courtship is, that it is Dr. M.'s first love, and that he would not wait, so the lady said yes! As you have seen this lady, I would give you this piece of news.

*   *   *   *   *

I beg you will give a true account of my dear niece's and the children's health, not forgetting the babe and how she will be named, that I may enter the same in my biographical account.

I remain, my dear nephew,
Your most affectionate aunt,
Car. Herschel.

The second Day-Book concludes in July, 1839, and is in all respects like the preceding one, but contains here and there touches and sentiments of which her own words can only do justice.

Aug. 3rd.—I went to buy some clothing for wearing at home, and went to my mantua-maker to give directions. I had to climb up to the third story, and I was of course quite knocked up when I came home, but it is my intention

* How d'ye do?   † Betrothed.

to continue to take some exercise as long as the weather and the length of the afternoon will permit.

Aug. 26*th.*—My niece Knipping came in the afternoon to assist me in some needlework—we did not do much!

Sept. 25*th.*—To-day I was made happy by a visit of Alexander Humboldt; which, though it was extended to the utmost limit of the time which this interesting man could spare me, was too short for all I wished to hear and had to say, which, as the theme of our conversation was my nephew, may be easily imagined.

Oct. 5*th* & 6*th.*—Mr. Hohembaum and the carpenter were with me to pack up the seven-foot telescope. I assisted as well as I could, being very ill all the while.

Oct. 7*th.*—Dr. G. called for a moment, but nobody else!

Dec. 10*th.*—I went in the evening to the concert, where I exposed myself most sadly by falling a-crying when the King most kindly came to me to inquire after my health. I do not think I shall have the courage to show myself there again in a hurry.

Jan. 27*th.*—This is the first day since the 30th December that the ice is detached from my sitting-room window.

Jan. 31*st.*--Mr. Hausmann brought me some Journals, and talked for an hour of old times with me, as he ever does, good man!

Feb. 7*th.*—A letter from my niece came this morning by the Hamburger post, which will make me happy for some time, and make me bear my painful solitude more patiently.

March 17*th.*—Thank God the 7th and 16th March are got over, and I begin to recollect that I have much else to do than bewail myself at being obliged to spend such days severed from all that *are,* or *were, so dear!* . . . . I found my poor friend [Mrs. Beckedorff] very much altered, but before I left her I thought she looked a twelvemonth younger for our two hours' chat. But we both were obliged

to part, for we could no more. Yesterday she sent me
some fine flowers, as usual on my birthday. Dr. Mühry
left a card ; two of my nieces called, and Hofräthin
Ubelode brought me some flowers. They left me fatigued
to death, to spend the long evening in solitude.

June 18th.—Yesterday Mr. Hausmann came to see me,
and brought the Philosophical Magazine for June, in which
I had the pleasure to see that Dr. Lamont has observed
three of the Georgium Sidus satellites.

July 3rd.—Dr. G. brought me an extract from *The Sun*
that my nephew has been created a baronet on the occasion
of the coronation.

July 9th.—My nephew arrived in Hanover in the evening.

July 10th.—In the afternoon I saw him and my little
grand-nephew for a few hours.

July 25th.—My nephew and his son took tea with me,
and we soon parted, without taking leave, and next morning
I am told they left Hanover at four in the morning. More
I cannot say !

Oct. 24th.—Mr. Hausmann came in the forenoon and
took the box with the mirror of my sweeper with him, and
in the evening he came to receive the stand. I am glad my
poor sweeper is now in good hands !

Oct. 29th.— Mrs. Knipping* spent an hour with me in the
dusk of the evening, and read an act of a play.

Dec. 30th.—In the afternoon Fräulein S. came to see me,
but she is deaf. I talked with her for a couple of hours
without either of us being the wiser.

Jan. 5th.—Went in the evening to the concert ; had some
talk with the Levies, who delighted the company with their
performance, especially the youngest son, eight years of age,

---

* Mrs. Knipping was the daughter of Dietrich Herschel, and the one of
the family Miss Herschel loved.

X

who gave several pieces on the French horn. Conversed with several persons besides the Prince Solms.

Jan. 20*th.*—I have been to the concert last night to hear the wonderful violinist, Ole Bull. It was very crowded for the confined room, though the largest in Hanover next the play-house. By the help of Miss B. and the M.'s I got safely through the crowd to my chair. But I was somewhat disappointed, for, by the report of those who had heard Ole Bull before, I expected to hear a virtuoso on the violin who would have given us an idea of the manner of performance of a Jordine, Kramer, Jacob Herschel, and Dietrich too; but it is more like conjuration than playing on a violin.

Feb. 12*th.*—Dr. Lissing paid me a visit. He wished me to subscribe to a work on Magnetism, but I think it would look only like affectation to let my name appear among the learned subscribers on a subject of which I know so little.

March 16*th.*—Mrs. Beckedorff sent me two beautiful flowers, accompanied by her good wishes, which she never forgets to do on my birthday. Mde. Knipping, and others, came to wish me to live many more years,—but what can I say?

March 23*rd.*—I was at the last subscription concert. His Majesty was there, and asked me how I did? I said, tolerably! This was all our conversation.

July 16*th.*—The whole of yesterday I had no other prospect but that it would have been the last of the days of sorrow, trouble, and disappointment I have spent from the moment I had any recollection of my existence, which is from between my third and fourth year. . . . . In the night I fell out of one fainting fit into another, and when I came to my recollection, between six and seven in the morning, I found Dr. G. sitting before me talking loud in his usual nonsensical way. Him had Betty called in her fright, for his wife (who is of use to nobody) is gone to spend the

summer months in the country. Mde. Knipping also is
away.

July 25th.—Mr. Hausmann, junior, and Mr. Hohenbaum
called to look at the photographical drawing. I am told it
is the only specimen of the kind in Hanover.

This Day-book, No. 2, is now full, and I shall not be easy
till it is deposited in a portfolio, in which will also be
found the Mem.-book 9. . . . . It often enables me to con-
tradict erroneous impertinent notions concerning my brother
William's disinterested character.

I am *now* not able even to look over, much less to correct,
what I have scribbled, but it must go as it is. Perhaps my
dear niece may look into them at some leisure moment, and
she will see what a solitary and useless life I have led these
seventeen years, all owing to not finding Hanover, nor *anyone* in it, like what I left, when the best of brothers took
me with him to England in August, 1772!

SIR J. F. W. HERSCHEL TO MISS HERSCHEL.

SLOUGH, Oct. 23, 1839.

DEAR AUNT,—

. . . . Now let me reply to your two letters of
August 26 and October 10, the last of which, being so
entirely in your old style, made us very happy. I now go
so little to London, and then only on the business of the
Royal Society respecting this magnetic expedition, that it
has not yet been practicable for me to call on Dr. Küper,
whom I well remember, however, at Cumberland Lodge,
and since.

As to sending either of our boys to Germany, it is time
enough, as W. is yet only six years old, and I assure you he
is now learning German very fast.

M. desires me to tell you, in answer to your question
whether she preserves your letters, that she does so, mots

x 2

carefully. She is sorry she omitted saying so in her last, in which she replied to everything else. So do I, you may be sure.

The Fables arrived safe, and W. must thank you for them himself, as well as for your care of him in Hanover.

I had the honour to meet at dinner, at Sir Gore Ouseley's, the other day, H.R.H. the Duke of Cambridge. He was very particular in his enquiries after you. He is quite well, and his affable and agreeable manners make him generally beloved.

Your letter of October 10th relieved us of much uneasiness, after the alarming account with which the former one was filled. When you once more begin to write about *die Gelehrten*, &c., I know all is well. So God bless you, and believe me,

Dear aunt, your affectionate nephew,

J. F. W. HERSCHEL.

MISS HERSCHEL TO LADY HERSCHEL.

Jan. 10, 1840.

MY DEAREST NIECE,—

*   *   *   *   *

Perhaps you may have heard that in the early part of its [the forty-foot telescope's] existence, "God save the King" was sung in it by the whole company, who got up from dinner and went into the tube, among the rest two Misses Stows, the one a famous pianoforte player, some of the Griesbachs, who accompanied on the oboe, or any instrument they could get hold of, and I, you will easily imagine, was one of the nimblest and foremost to get in and out of the tube. But now!—lack-a-day!—I can hardly cross the room without help. But what of that? Dorcas, in the *Beggar's Opera*, says, "One cannot eat one's cake and have it too!" . . . .

I will only thank you once more for your charming letter, and beg to be kindly remembered to all who are dear to you, and to give an embrace extraordinary to the dear little ones around you, and not forgetting to include my *dear* nephew in the general hug! and believe me,

My dearest niece,

Yours and his most affectionate aunt,

CAR. HERSCHEL.

P.S.—One anecdote of the old tube (if you have not heard it) I must give you. Before the optical parts were finished, many visitors had the curiosity to walk through it, among the rest King George III., and the Archbishop of Canterbury, following the King, and finding it difficult to proceed, the King turned to give him the hand, saying, " Come, my Lord Bishop, I will show you the way to Heaven!"

This was in the year 1787, August 17th, when the King and Queen, the Duke of York and some of the Princesses were of the company.

I hope the book where the visitors were noted, has been preserved? Some time after it was kept by other hands; but before I parted with it, I copied some pages which put me sometimes in mind of persons who were interesting to me.

These scribblings will come to you among the rest of my scraps. Good-bye!

MISS HERSCHEL TO LADY HERSCHEL.

HANOVER, *Jan.* 10, 1840.

MY DEAREST NIECE,—

*   *   *   *   *

. . . . For the last month past I have been so much disturbed and fatigued by visitors who came to wish me a happy New-year, &c., for I have of late gained the

acquaintance of half a dozen ladies, added to two who were
in the habit of visiting me between the hours of twelve at
noon and six or seven in the evening; [for the first two or
three hours, after having passed a sleepless night, I am
obliged to spend in the manner as perhaps you may have
seen Lord Ogleby did in *The Clandestine Marriage*].

But now, from seven to eight till between eleven and
twelve, I am left to amuse myself as well as I may, but it is
no easy task to turn books into companions by one who has
no eyes left; but there is no help for it.   There is neither
man, woman, nor child in Hanover to be found but they
must spend the evening at balls, plays, routs, clubs, &c.,
and not a month goes over one's head without a jubilee
being celebrated at enormous expense to someone who has
fifty years enjoyed title and salaries for doing his duty (any-
how, perhaps).

But what a contrast between a jubilee auf der Börse* at
Hanover and the one at Slough, † described in your letter
with which I was made happy January 4th.   The company
so select—for I figure to myself none but angels from above
were listening to, and joining their kindred in the chorus
below! . . . .   Before I take leave of this jubilee I must
beg the excellent poet of the song to accept my hearty
thanks for remembering me so kindly in verse 4, and for not
letting the poor forty-foot telescope‡ depart in silence.

---

* On the Exchange.

† The whole family party assembled at Christmas in the tube of the great
telescope, and sang a ballad composed for the occasion.

‡ " The telescope, as you know, is laid on three stone piers horizontally.
It will be fresh painted to-morrow, and afterwards every three or four years,
as it wants it, and it looks very well.   The observatory will remain nearly as
it is.   The apparatus of the telescope is *inside of the tube*, and will be riveted
up from all intruders.   And all the polishing apparatus is *fixed* on the spot."
—*Letter of Sir John Herschel, Feb.* 28, 1840.

The great mirror is now put up in the hall of the house—"Herschels"—at

MISS HERSCHEL TO LADY HERSCHEL.

April 5, 1840.

My dearest Niece!—

Your delightful letter of March 8th, which I received
about a week after that of my dear nephew, could never
have come at a more needful time for chasing away the
melancholy impressions my friends' losses and misfortunes
have had on my spirits. On the 7th of March Dr. Mühry
came to wish me joy on my nephew's birthday. Nine days
after, when they all used to come and bring me flowers,
&c., the whole family were thrown into despair by the death
of Dr. C. M., who died by his own hands (thirty-four
years old). About a week before I had spent an evening
with him at his grandmother's, when he begged me to thank
my nephew once more for giving him a letter of introduction
to Dr. ——, at Oxford. This poor man was spoiled by
being made too much of from his infancy. As a boy of
seven or eight, he was brought to England to visit his
grandmother and aunt, and was loaded with costly presents
by the Princesses, and fed with nothing but dainties, till,
when grown up, nothing but what was most extravagant
would satisfy him. The 30th of March our friend P——
was buried, eighty-three years old. On my birthday a circular letter came by post, announcing Dr. Olbers's death.
So, I must say once more, my nephew's and your dear letter
came very seasonably to turn my thoughts to something
more cheering. . . . .

Now I am in two minds whether I shall turn to my dear
niece or have done with you first. But out with it! I would,
if you have no objection, draw on Mr. Drummond for £52,

Slough, by the present tenant, Mr. Montressor, who has spared no pains
to do.honour to the relics as well as to keep up the character of the old
fashioned "habitation," which owes much to the taste and judgment he has
bestowed on it.

*   *   *   *   *

and if I should (as it seems) live to the age of Methusalem, come again for the same sum after the 10th of October next. For this is quite enough for me to live with credit, and more would only be a trouble to me.

I am tired, and can write no more just now, but for our amusement I will, some time or other, give you the history of the few days you were in Hanover, in July, 1838. For all that past was like Sheridan's *Chapter of Accidents*. If I could only have had a few hours' of private conversation with you then, much trouble would since have been spared me.

I hope to have soon some account of how your new situation agrees both with papa, mamma, and the little bodies. How many English miles is it from London?

. . . . My sweeper, which I should have been so happy to put in the hands of my little grand-nephew, and teach him to catch comets till he could do something better (O! why did I leave England!) is now in the hands of the good, honest creature, Director Hausmann, and the seven-foot telescope is also saved from being sold for an old song. . . .

MISS HERSCHEL TO SIR J. F. W. HERSCHEL, BART.

July 6, 1840.

*   *   *   *   *

But at another time, when perhaps I may find myself a little better, I will amuse my dear niece with introducing some of my acquaintances to her notice. Some of the family of General Halkett,* at least, she will not be displeased at knowing personally. Last night the sister of the general, Mrs. W. Clarke,† a widow, sat an hour with

---

* General Baron Hugh Halkett, a distinguished officer of the German Legion, died 1863.

† Miss Herschel gave special directions that, after her death, her snuff-box should be given to this lady.

me, and said she would next summer visit her late husband's relations in England, and then she would not fail of seeing you. You must love her for my sake, for she really takes some pains to give me pleasure, bringing me flowers, taking me an airing in her fine English equipage, &c. I must not forget the general's lady, a second wife, of course a step-mother of my young friend. She is Scotch (a Graham), and brought me little Christmas pies in her reticule on New-year's Day, of the young lady's making—the only good kind I have tasted in Hanover, and they were as good as my nephew's mamma ever made.

MISS HERSCHEL TO LADY HERSCHEL.

August 3, 1840.

MY DEAREST NIECE,—

. . . . But first and foremost, I must beg you will give my best thanks to my dear niece Caroline for her very sensible and very clever letter, and I only wish I may be often favoured by her fair hands with such favourable accounts of all your health and contentment with your new situation.

I am not able to write long letters, and must content my-self with saying, in as few words as possible, that if my nephew thought the seven-foot telescope worth the acceptance of the Royal Astronomical Society, it is well! . . . . (Mem.—Its only being painted deal was, because it should look like the one with which the Georgium Sidus was discovered.)

I have also the proceedings of the Royal Irish Academy to thank you for, twenty pages. I suppose I have nothing to do but to accept them. But I think almost it is mocking me to look upon me as a Member of an Academy; I that have lived these eighteen years (against my will and inten-tion) without finding as much as a single comet. But no more of these terrible eighteen years just now. . . . .

My dear nephew, if I did not feel myself seriously declining very fast, I would not incommode you at present
(when your time must be so precious) with such letters as
my two or three last have been.

But going many nights to bed without the hope of seeing
another day, I think it my duty to guard you against putting any trust or confidence in ——. He and the whole
family have never been of the least use to me ; and for all
the good I have lavished on them, they never came to look
after me, but when they had some design upon me.

In short, I find that all along I have been taken for an
idiot, or that at least I am now reckoned to be in my
dotage, and therefore ought not to be mistress of my own
actions.   But, thank God, I have yet sense enough left to
caution you against being imposed upon by a stupid being
who would make you believe I died under obligations to any
of the family.   I know he has already, without asking my
leave, passed himself off for my guardian, and is vexed at my
being able to do without him.   But I could not live without
that little business of keeping my accounts ; and by my last
book of expenses and receipts may be seen, that I owe
nothing to anybody, but to my dear nephew many many
thanks for fulfilling his father's wishes, by paying for so
many years the *ample* annuity he left me.

SIR J. F. W. HERSCHEL TO MISS HERSCHEL.

August 10, 1840.

. . . . The telescopes are now, I trust, properly
disposed of.   Mr. Hausmann (who will value it) has the
sweeper.   The five-foot Newtonian reflector is in the hands
of the Royal Astronomical Society, and will be preserved
by it as the little telescope of Newton is by the Royal
Society, long after I and all the little ones are dead and
gone.

MISS HERSCHEL TO LADY HERSCHEL.

Dec. 27, 1840.

. . . . There is another circumstance on which
account I feel not very easy, which is that by leaving Slough
you are separated from all your usual friends, &c., doctors
and all; but pray keep up your spirits, for the days are
already a cock's stride longer, and my windows have now
been covered with ice for the last three weeks, which is long
enough in conscience; therefore I hope to see a change
every morning when I can get my eyes open, which is never
the case till near eleven o'clock.

There have been some English gentlemen with Mrs.
Beckedorff on business, who, in conversation, among the
rest, were saying that the keeping Christmas in the German fashion was coming to be very general in England;
but I hope they will never go such lengths in foolery as
they do here. The tradespeople have been for many
weeks in full employ framing and mounting the em-
broideries of the ladies and girls of all classes, for there
exists not a folly or extravagancy among the great but it is
imitated by the little. The shops are beautifully lit up by
gas, and the last three days before Christmas all that
could be tempting was exhibited in the market places in
booths lighted up in the evening, where all run to gaze
and get a liking to all they see. Cooks and housemaids
present one another with knitted bags and purses, the
cobbler's daughter embroidered neck-cushions for her friend
the butcher's daughter, which are made up by the uphol-
sterer at great expense, lined with white satin, the upper
part, on which the back is to rest, is worked with gold,
silver, and pearls.

But I find too much difficulty to write in these short
days, else I could write a book about the nonsense which

is going on in this city. I have for this last month been completely tired out with this Christmas bustle; but now the balls at the Bourse, given by the shopmen to the daughters of their masters, will be succeeded by the masquerades in Lent, an amusement which in the good old times was only for the nobility, but from which they are now excluded. . . . .

MISS HERSCHEL TO SIR JOHN HERSCHEL.

Sept. 1, 1840.

. . . . I owe you many thanks for relieving me two whole days sooner from the anxiety of having been misunderstood by you, and now I am happy, and *all is well!* But there are times when I should like to have some talk with you or my dear niece, to put you in mind of many past events, but if you will excuse the style and the spelling, &c., &c., on account of my eyesight being so uncertain, I will at times try to amuse you with what passed in old times, for my memory is as good as ever [this is in her ninety-first year]. (N.B.—Year of the past.) Writing this, puts me in mind that I never could remember the multiplication table, but was obliged to carry always a copy of it about me.

SIR J. F. W. HERSCHEL TO MISS HERSCHEL.

August 10, 1840.

. . . . Did I ever tell you that I had lately brought together the observations of four or five years, proving beyond all doubt *a* Orionis to be both a variable and a periodical star, and one of the most remarkable among them? Its period is about a year, and it changes in that time from a lustre superior on some occasions even to Rigel, to a degree of brightness nearly on a par with Aldebaran.

MISS HERSCHEL TO SIR J. F. W. HERSCHEL.

Feb. 24, 1841.

*   *   *   *   *

I intended to have made some remarks to you about several things which are said in those pages which came enclosed in the letter of February 3rd. I suppose it is not expected to acknowledge the receipt thereof, but if there is anybody to whom my thanks are due, I beg you will do it for me, because I am not capable of writing to strangers. But to you I cannot help pointing out several things which displease me very much. . . . .

I think whoever reads the Preface to the description of the forty-foot telescope (see " Philosophical Transactions," June 11, 1795), would not accuse him of jealousy— which also may be seen by the four volumes on the construction of Specula, which your father left behind in MSS., (to which you added those excellent drawings of the machinery, &c.), which it was my care, for half a dozen years at least, to save them from being devoured by the mice, by placing them on a table in the middle of the library, where I was obliged to leave them when I left Slough, for I could not find a better place for them.

Your father was latterly most miserably stinted for room, and I fear many, many things have met with destruction in consequence of being put by in corners among rubbish when not in use. For instance, when polishing and the foci were to be tried, by three apertures [tubes], which generally wanted to be repaired first; (for the twenty-foot they were made of pasteboard, but for the forty-foot of light deal) and I was directed to hold them before the mirror, and, listening to the report of the trial, was glad to hear " All right, three foci perfectly alike ! " and the work proceeded to perfect the polish. Dear nephew, I stick fast, and must give over talking about these things; it downright fatigues me. But these folks would not have called the Herschelian construction useless if they had seen the struggle, during the years from 1781 to '86, to get a sight of the Satellites of the Georgium Sidus, when, after throwing aside the speculum, they stood broad before us. . . . . . Pray, does South live still ?

MISS HERSCHEL TO SIR J. F. W. HERSCHEL.

March 31, 1841.

*   *   *   *   *

Not to send blank paper, I will fill it by copying from my Day-book the names of the visitors I had to receive on the 16th of March.   This I can do mechanically and by feeling, and it serves to pass away the time, as I cannot see to read for any length of time.

By way of being ready to see anybody by twelve o'clock, I rose an hour earlier than usual, but before I was dressed, Mrs. Beckedorff and Mrs. W. Clarke sent each a beautiful moss-rose and card.   Soon after, Mrs. Clarke and General Halkett came; Generalin Borse and daughter brought violets; Frau von Both; Ober Medicinal-Rath Mühry; Miss Beckedorff; Madam Groskopff; Hofräthin Ubelode brought mignonette; Oberjustiz-Rath von Werloff sent crocuses ; Fräulein von Werloff sent a card and hyacinths ; Dr. Groskopff, Hauptman Buse, Alexis Richter, Major Müller ;—all these I saw between twelve and four o'clock, and several for a good while together.   I talked and complimented myself into a fever, of course "looked blooming," and am to live to be a hundred years old.   What stuff ! After eating my solitary dinner I tried to get a little sleep, as I generally do, but before I could compose myself enough, two of Major Müller's sisters came and remained two hours with me ; after they left me, Fraulein von Werloff sent her companion, a Mademoiselle H., and a sister, to keep me company till ten o'clock. With difficulty, and the help of Betty, I got into bed, but could get no sleep, nor the whole day after.

MISS HERSCHEL TO LADY HERSCHEL.

HANOVER, *July* 31, 1841.

MY DEAREST NIECE,—

If it was not that I ought to thank you for your kind letter of June 9th, I should perhaps not have now the spirit to take up the pen; but your letters always, especially the last, contain, besides the many consoling expressions, such very interesting information, that I would not for the world risk to lose the monthly sight of your dear handwriting, by omitting to return at least my grateful thanks for your kind communications of what the present philosophers are about.

I think I can form some idea of the author of the book on philosophy (and godfather of our little Amelia), from what I recollect to have read some years past in some quarterly publication by a Mr. Newell, in defence of Sir Isaac Newton. In short, it met with *my approbation!* There is for you! What do you say to that?

I do not wish to write in what my dear brother William used to call a Dick Doleful style, when our brother Alexander was in the dismals, and out of which we often succeeded in laughing him. But I cannot just now turn to anything of a cheering nature, for yesterday, the 30th, our Queen died, and I have been very unwell in consequence of the violent change in the weather. . . . .

The following letter refers to the intended removal of Sir J. Herschel and his family to Collingwood, which he had purchased :—

MISS HERSCHEL TO LADY HERSCHEL.

HANOVER, *August* 2, 1841.

MY DEAREST NIECE,—

. . . . I could wish to know something more about the place where you now are.* How many miles is Collingwood from London? How many from Hastings? Have you any good people or neighbours about you? I think I read in Watson's Gazetteer, Hawkhurst to be full of poor, and, what is worse, of smugglers. Pray take care of the dear boys and children, that they are not kidnapped in their little rambles from home.

I can for the present only say so much of myself that my friends are almost going to kill me with their visits, like, as they say, the cat did her kitten with kindness. On Sunday I was even honoured with a visit from the Duchess of Anhalt Dessau and the Princess of Rudolstadt—the latter a little astronomer—who remained a whole hour with me. They are both daughters of the late Queen.

MISS HERSCHEL TO LADY HERSCHEL.

HANOVER, *Feb.* 3, 1842.

. . . . Your mentioning the Government gift of the Kew Observatory to the Royal Society, recalls to my mind the struggles through a life of privations during the lapse of between twenty and thirty years, till my brother had realised a capital sufficient for living in a respectable manner by making seven, ten, twenty, and twenty-five-foot telescopes. For it was in 1782 when Mr. De Mainborg, the King's private astronomer (formerly one of his tutors) at Kew, died, and my brother, in consequence of the discovery

---

* The family of Sir J. Herschel had left Slough and settled at Collingwood, near Hawkhurst, Kent, now the family residence.

of the G. Sidus, was called from his lucrative employment
at Bath.   His friends had no other idea but that he was to
succeed Mr. De Mainborg at Kew.   But it was otherwise
decreed, for the King was surrounded by some *wiseacres* who
knew how to bargain, and even £100 were offered if he
would go to Hanover!

But you know by what I once wrote on a former occasion
that he settled at Datchet with £200 per annum, after four
months' travelling between London, Greenwich, and Windsor, and moving his workshop and instruments from a house
at Bath, of which he had a lease.   And at Michaelmas,
1782, was the first £50 he ever saw of the King's money.
This happened at the time when Parliament had granted to
the King £80,000 a-year for encouraging sciences.   This I
only knew by what I heard at that time, and that Mr. West,
R.A., with his giant Judas, Jervis, who made the altar-piece
for St. George's chapel (which I once heard Mrs. Beckedorff
say had cost the King £30,000), and Herschel, were the first
who benefited by this grant.

I am full of expectation of W.'s promised description of
the Christmas entertainment; but put him in mind that I
do not understand Latin.   Of A's Greek, I think I can be a
judge, knowing the letters of the alphabet in consequence of
their being used in the astronomical catalogues. I
hope music is still in favour with the family; often I lament
that at the time of our quitting Bath in such a hurry my
brother's musical treasures were scattered, and given to the
winds.   Among the rest there was a song for four voices,
" In thee I bear so dear a part," which was just going to be
published by desire, for it was sung by the first performers
from the London theatres, and encored, between the acts of
the oratorios.   I wrote it out ready in parts during my
brother's absence ; but he could not find a moment to send
it off, nor to answer the printer's letters.

Oh! how I should like to hear some of the glees and catches sung by the great and little family in the music-room at Collingwood; but it was not to be! and I had rather leave off and leave some room for the many good wishes to yourself, my dear nephew, and all those who are dear to you, and believe me,

My dear niece,

Ever your most affectionate aunt,

CAROLINE HERSCHEL.

MISS HERSCHEL TO LADY HERSCHEL.

HANOVER, *March* 3, 1842.

MY DEAREST NIECE,—

. . . . Nothing runs in my head but what concerns my family and connections, and I am at present living over again the last eighty-nine years of my existence. . . . . But I will leave off teazing you with these old stories with which I am obliged to amuse myself, for I cannot see to work or read, and must therefore either sleep or scribble, for my visitors come mostly in the forenoon, their evenings being taken up with public amusements or private parties, of which I have not been able to be a partaker these three years, for I see by my account-books it is so many since I left off subscribing to the play. But to please Mrs. Clarke I made the experiment on the 3rd of February, whether I should come home alive after seeing *King Charles II. in Wapping*, acted at the English Ambassador's. Mrs. Clarke came about twelve with an invitation from the Honourable Mrs. Edgecombe—their house not containing a room large enough for giving great balls, they contrived this way of entertaining the company. The enclosed playbill will show the rest.

There was no time for consulting milliners, and Mrs.

Clarke assisted me in looking out something from what I had worn some years back, cap and all. (N.B.—The latter of my own making.) I must give you here a German saying, if you do not know it, which is, "Einen jeden Narren gefällt seine eigene Cappe!"* but I cannot say that I was much pleased with mine, I have so very few grey hairs left, which, however, I was told were much admired!

Mrs. C. left me with a promise of sending her chair and servant at three-quarters past seven, and was waiting in an ante-room for me to assist me in getting further, and, indeed, the whole evening she did not withdraw her arm from me till she had put me in my chair again, and the next morning she was with me almost before I was out of bed. The King, Princess of Rüdolstadt, and one of the Princes of Solms were among the company, and I did not come home without receiving their notice. But I shall not venture on such pranks again, I promise you!

* * * * *

As I am writing this I see it will be my birthday, when I shall be ninety-two years, if I live. My nephew's is the 7th, and he will be fifty, but for all that do not think him to be an old man. His father was fifty-four when he first saw the light. . . . .

The King of Prussia left magnificent presents among the courtiers, and Generalin Halkett was here on Sunday, and promised to bring me a snuff-box to look at, which the general has received. I begged she would not, for the ladies wear no pockets, and lose their purses, &c., as I daily hear by the town crier. Their pocketkerchiefs they carry open in their hands, which I think very indelicate; I daresay it is not the fashion in England. . . . .

. . . . I would not wish on any account to see either my

---

* Every fool is pleased with his own cap.

Y 2

nephew or you, my dear niece, again *in this world, for I could not bear the pain of parting once more;* but I trust I shall find and know you in the next. And as long as I can hold a pen, let us, I beg, commune with one another by letter!

MISS HERSCHEL TO LADY HERSCHEL.

HANOVER, *June* 2, 1842.

MY DEAREST NIECE,—

    A thousand thanks for your kind letter, which contains ever so much comfortable and satisfactory information, such as heart can but wish. . . . .

I have begun a piece of work which I despair of finishing before my eyesight and life will leave me in the lurch. You will perhaps wonder what such a thing as I may pretend to do [can be], but I cannot help it, and shall not rest till I have wrote the History of the Herschels. I began, of course, with my father and his parents. My father was born in January, 1707, and I have now only got so far as the beginning of 1758, and it begins to interest me much, but I doubt whether I shall live to finish it, but think it a pity it should be thrown away.* . . . .

    . . . . Do not forget to thank my little nephew for his pretty letter. His description of the method his papa makes use of in teaching mathematical figures, I prefer to that of his grandfather. He used, when making me, a

---

* In answer to this announcement her niece wrote : " Herschel bids me say he is quite delighted at the idea of your undertaking the family history, but he insists upon it that you prove his descent from *Hercules,* and I dare say in this age of relics, we could contrive to find in the rummaging of old traps turned out at Slough, a veritable piece of the old *club* which has by fortunate accident served as part of the ladders of the forty-foot telescope ! or perhaps you remember its slipping down the mouth of the great telescope one night when it was turned in the direction of your ancestor's constellation, as a sign that he confessed himself *outshone* by your *labours.*"

grown woman, acquainted with them, to make me sometimes
fall short at dinner if I did not guess the angle right of the
piece of pudding I was helping myself to!

MISS HERSCHEL TO LADY HERSCHEL.

HANOVER, *July* 7, 1842.

MY DEAREST NIECE,—

I have just now been reading your dear letter of June
7th once again, but I shall take care not to look into it for
yet a while, else I run the risk of going mad when thinking
of my running away from a country where I might have
been an eye-witness, and sometimes a partaker, of so much
domestic happiness. But it is no matter now, and of no
use fretting about it; I am only sorry I cannot go on with
my history as fast as I could wish, for I feel too unwell to
be doing *any* thing for any length of time. . . . .

. . . . I am glad my dear nephew finds pleasure in
giving up so much of his valuable time to his dear sons;
for my hair stands at an end on hearing what beings are
continually expelled from *our* Eton here, all owing to
ignorant ambitious parents trusting entirely to unprincipled
hirelings.

Though my poor brother seemed to have no hands in the
education of his only son, I know, from having been present
at many private conversations he had with Dr. Gretton, that
nothing was done without his approbation and advice.

. . . . The "Astronom. Nachrichten" have latterly
been filled with tables and too much mathematic (for me).
The last numbers, 450, 451, contain an account, by Struve,
of the purchase of Olbers' books, &c., for the library of the
Observatory at Pultowa. This puts one in mind of Olbers
saying somewhere, I had discovered five comets. Who
wanted him to give the number of *my* comets when he

knew them no better ?   As far as I recollect, Dr. Maskelyne
has observed them all, and his observations on them are, I
daresay, all printed in the volumes of the Greenwich Observations—at least of some he has shown me the proof sheets.
I never called a comet mine till several post days were
passed without any account of them coming to hand.   And
after all, it is only like the children's game, " Wer am ersten
kick ruft, soll den Apfel haben !   Wo sie denn alle rufen
kick ! kick ! und so,"* &c., &c.

I long for the return of the messenger, for I heard to-day
that Bessel and Encke were gone to the philosophical
meeting in England, and I expect to hear a great deal of
news.   But first and foremost I wish to see in your next
that yourself and my dear nephew, with all the dear *little,
little* ones, continue to be well and happy. . . . .

P.S.—My head is full of my History, and I go on but
slowly, because I cannot sit up for any length of time.   I
am only at my fourteeth year, and have just parted from my
brother, William Herschel I., who is returned after a fourteen nights' visit to us, to England, Leeds in Yorkshire
(where he must be left for some time), and I cannot go on
till I have recovered from the parting scene.

You remember, you take the work in whatever state I may
leave it, and make the best of it at your leisure.   Adieu.

TO SIR J. F. W. HERSCHEL.

HANOVER, *August* 4, 1842.

M͞Y DEAREST NEPHEW,—

. . . . Major Müller is not yet returned, and is not
expected till September, from his measuring business, and
besides him there is not one astronomer, or, I may say,
rational man in Hanover to whom I could apply for infor-

* He who first cries "Kick !" shall have the apple.

mation in matters which are above my understanding. But
in my next I hope to say more, or rather a great deal about
your "Chrysotype," for I had a visit to-day from a Berg-
Rath-W., who seems to be much interested in these discoveries. . . . . How I envy you having seen Bessel—the
man who found *us* the parallax of 61 *v* Cygni. . . . .

. . . . I believe I have water on my brains, and all my
bones ache so that I can hardly crawl ; and besides sometimes
a whole week passes without anybody coming near me, till
they stumble on a paragraph in the newspaper of Grüthou-
sen's discoveries, or Lord Queenstown's great telescope,
which *shall* beat Sir William Herschel's all to nothing, and
such a visit sometimes makes me merry for a whole day.

SIR J. F. W. HERSCHEL TO MISS HERSCHEL.

COLLINGWOOD, *Aug.* 9, 1842.

MY DEAR AUNT,—

M. tells me I must finish this letter with an account
of the total eclipse of the sun seen at Pavia by Mr. Baily,
and at Turin by Mr. Airy. At Pavia it was very finely seen,
and as soon as the sun was totally covered, the dark moon
was seen to be surrounded with a *glory*, like the heads of
saints in old pictures. While he was admiring this, a great
shout from all the population of Pavia broke out at once,
which was caused by the sudden appearance of three purple
or lilac-coloured flames, which seemed to break out from
the edge of the moon. At Milan the same was seen, and
the people shouted out "Es leben die Astronomen!"* as
soon as they saw the flames.

I am glad you got my Chrysotype pictures safe. The
present beautiful sunshine has given me an opportunity to
make great progress in photography, and the enclosed pho

* The astronomers for ever !

tographic copy of a little engraving or two may serve to amuse you. Meanwhile the star reductions are not forgotten. Thirty more sweeps only remain to be reduced, and I am already in the engraver's hands with the nebulæ pictures. And so the world wags with

Your affectionate nephew,

J. F. W. Herschel.

SIR J. F. W. HERSCHEL TO MISS HERSCHEL.

. . . . On the 30th of last month I finished the *reductions* of all my Cape nebulæ and double stars, and have got all the former and all but a very small number of the latter arranged in catalogues in order of Rt. Ascension for the epoch 1830, January 1st. Thus these two most important parts of my Cape work are at last secured against *loss*, and it will not be long now before I shall begin to prepare for the work of publication in good earnest. I mean as to the narrative part.

Dec. 8, 1842.

LADY HERSCHEL TO MISS HERSCHEL.

Jan. 12, 1843.

. . . . Your nephew sends you his translation of Schiller's beautiful and *instructive* poem, " The Walk," in which he tied himself down to the original metre, and each couplet contains the sense of the corresponding couplet in German, so that the full strength of the English language was required to do justice to the comprehensiveness of Schiller's ideas. There was a beautiful walk up the side of Table Mountain which always reminded Herschel of this poem, and made him love it ; and lately there have appeared in an Edinburgh Review translations of all Schiller's minor poems, some of which are well done ; but he thought " The

Walk" deserved to be better rendered, so he set about it, and distributed it among his friends as his Christmas sugarplum. The number of interesting autographs, criticisms, witticisms, &c., which have been thereupon returned, will make an amusing packet. One lady says (alluding to the singularity of the hexameter in English) that she found it difficult to get into the *step* of the Walk; another, that the *Walk* had got into a *Run*, it was so often carried off by friends from his table; another, not knowing whence it came, intended sending it to Herschel for his opinion on its merits! another, while admiring the ideas, says "to the *verse* I am *averse.*" The good Misses Baillie, of Hampstead, have been greatly delighted with it. They desired their kindest remembrances to you.

*   *   *   *   *

MISS HERSCHEL TO LADY HERSCHEL.

HANOVER, *March* 1, 1843.

. . . . Nine o'clock in the evening (February 19). This is the first moment of quiet after six days in tumultuous joys by all living beings, from the most highest to the most lowest, and I will give you here an account of what share I have had in the rejoicings. In the first place, I must begin with confessing that I have been uncommonly ill of late, and nobody came near me to comfort me; for all my friends were too busy with gala-dresses, or else laid up with colds, &c., from shopping in bad weather, and paddling about in the snow, and I am at this moment ignorant of how they have fared. . . . .

I have not time to fill the paper, for my friends begin now to take up my little time of my *short* forenoons, and the evenings I cannot see; so here I send what I have been scribbling, and will only add that the enclosed programme was sent me, on the 14th by the Crown Prince, who having inquired through somebody after my health, and hearing I was well, and preparing for illumination, was much affected ; and yesterday his adjutant, Major Stolzenberg, brought me a message from the Crown Prince, including H. R. H. the Princess, with a present of their portraits.

      *   *   *   *   *

        *   *   *   *   *

TO SIR J. F. W. HERSCHEL, BART.

April, 1843.

My dearest Nephew !—

Many thanks for your dear letter, which I found on my breakfast-table on the morning of the 16th March,* . . . . when the Crown Prince and Princess were announced. Mrs. Clarke, who just came in, assisted me to entertain the royal and interesting pair for nearly an hour. They came in arm in arm, carrying an immense bouquet before them, which I heard afterwards they were returning with from the hothouses at Herrnhausen. As soon as the Princess was placed on the sofa, and I beside the same, the Crown Prince drew a chair close to me, chatting and joining in our conversation. I could not help giving the Princess the lines of your letter to read, where you mention them so prettily, and presenting her with " The Walk," which was lying among the flowers and the open letters before us on the table. It was a little rumpled in the coming, which she said made it the more welcome, as it would remind her of its having once been mine.

I intended to amuse you with the list of the names and titles of all the visitors I had to receive on that day, but you will find them one of these days in my Day-book ; and I will only say that it was rather too much to expect me to be

---

* Her 93rd birthday.

civil to upwards of thirty persons in one day, which lasted till evening, so that I had no time to eat a morsel, finding myself seriously ill.

*   *   *   *   *

May 4, 1843.

Memorandum for my next letter, made April 23rd.

To my Nephew. On reading your letter to the editor of the *Times,* of March 31st, I recollect having written down some observations of your father's on the zodiacal light; he never lost an opportunity of noticing anything remarkable during twilight, or in the absence of nebulæ, &c., and I remember also his explaining to me another kind of ray, which is after sun-setting, reaching up *very* high ; but this only appears for one or two nights at the equinox : but I have forgot all about it, and want only to speak here about a *temporary* Index to observations, in which I know a few of such-like memorandums were catalogued or carried in their separate books. With this Index your father was never satisfied, telling me, " I could not make an Index, it was a task Sir I. Newton had found too difficult to accomplish," . . . . and he would hardly allow me to make use of this book, after calling it a *temporary* Index. But it has often saved me a whole week's poring over the Journals for a memorandum. . . . .

MISS HERSCHEL TO LADY HERSCHEL.

June, 1843.

My dearest and best Niece,—

I must write a few lines by way of thanking you for your dear letter of May 9th. Your description of the splendid observations which are made on the roof of your own mansion, recall the many solitary and, at the same time, happy hours I spent on my little roof at Slough, when I was not wanted at the twenty-foot. And I cannot help at the same time regretting my having spent these last twenty years in so useless a manner, between roofs and houses which prevent my seeing even an eclipse of the moon when in a low part of the ecliptic, it passes away behind the houses of my opposite neighbours; and so did the glorious tail of your comet, of which, however, I have gathered all that has been said in the papers, besides what you and my dear nephew have been so kind as to communicate. . . . .

I have just been reading part of your dear packet over again, and am resolved to follow your advice, and say as little of what happens now as possibly I can help, and send herewith what I call the first part of my History, of which I wish you will in your very next give me your sincere opinion. I shall judge by it if I may go on, or lay down the pen for ever.

(I hope the packet containing my brother's biography has been safely taken care of among his papers, for I have no copy of it; pray let me know if you have seen such a packet, I think it is in quarto, and that I put it in a cover like all the MSS.)

Of the present I can only say that I have been unable to do anything beside keeping myself alive, and getting my clothes on by twelve at noon, so that I may be able to receive anybody who may call on me between that hour and eight in the evening.

*   *   *   *   *

This brings to my remembrance, that when I was god-mother to Mrs. Waterhouse's eldest sister in 1787, I was called away in the afternoon to help my brother to receive the Princesse Lamballe, who came with a numerous attendance to see the moon, &c. About a fortnight after, her head was off.

SIR J. F. W. HERSCHEL TO MISS HERSCHEL.

Collingwood, *Sept.* 13, 1843.

My dear Aunt,—

Again we are rejoiced by the sight of your hand-writing, and by the admirable and truly interesting History of your own younger days, which you have sent with your delightful letter, and which arrived perfectly safe, and, you may be sure, will be treasured as the apple of the eye, and often read and re-read. I began the reading of it last evening to all your grand-nephews and nieces who are old enough to understand it, and the History of their great-grandpapa's hardships after the Battle of Dettingen, and poor uncle Alexander's harsh treatment, and your own quiet, thoughtful activity and self-dependence, made on all my hearers, as well as on myself, an impression which I am sure will not easily be forgotten, and which I shall take care not to let them forget. We all entreat you to continue it, and you need not be in any fear about the *writing.* Your hand-writing (Gottlob,*) is still excellently good, and there was not a word either in your letter or in the " History" that gave me the least trouble to read. . . . .

. . . . I visited in London Mde. Taylor (whom you en-trusted with the pictures of your Royal visitors, which are very charming things, and seem as if they must be good likenesses). I did not find her husband at home, but she is a very pleasing person, and pleased me greatly by the respectful and friendly way in which she spoke of you. We hope to see them here, where they will be much valued, as will be the effigy or recollection of everybody that has been kind to you, or anything that has given you pleasure. . . .

The only news I have to send you is that of Capt. Ross's safe return with the South Polar Expedition after nearly four years' absence, having penetrated to the 79th degree of

* I thank God.

S. Lat., and discovered a new continent full of volcanoes and icy mountains, and the true position of the south magnetic pole. He anchored his ship upon the spot where the Americans say they found land, and found no bottom at six hundred fathoms!

MISS HERSCHEL TO SIR J. F. W. HERSCHEL.

June 4, 1844.

My dearest Nephew,—

. . . . For these last three months I have not been able to add a single line to my Memoir, but what you will find among my papers and memorandums; perhaps your daughter Isabella may, for her amusement some time or other, correct and write in the clear, my scribblings, for I find that in attempting to correct one blunder I am making two others in the same line. But I wish you might see, by what I say of myself, what trouble and invention it must have cost your father to enable me to assist in determining the places of all these objects, and I see with pleasure that your observations agree so nearly.

*   *   *   *   *

I was going to send, for the amusement of my dear niece, some description of what is going on here in Hanover, but I find it would be too much for my time and patience at present, and will only say that I believe they are all out of their senses.

There is an *Eisenbahn* * from Hanover to Braunschweig just now completed, which has turned them all wild. Some hundreds of high officers all (but the King) set off at eight o'clock to breakfast with the Braunschweigers, and returned with the same at three to dinner (eight hundred in number) in the orangery at Herrnhausen, from whence the Braunschweigers returned and were at home, I believe, again at eight.

Railway.

I am too tired at present, else I was going to tell you how
they are building. Hanover is now twice as large as when you
saw it last; nothing but castles will serve them any longer.
I have all this from hearsay, for I have not been downstairs
since February 3, 1842.

*   *   *   *   *

They talk of nothing here at the clubs but of the great
mirror and the great man who made it. I have but one
answer for all, which is, " Der Kerl ist ein Narr ! " * . . . .

MISS HERSCHEL TO LADY HERSCHEL.

March 4, 1845.

MY DEAREST NIECE,—

*   *   *   *   *

Have I understood you aright ? Saw you the ther-
mometer $1\frac{1}{2}°$ above zero ? the lowest I have heard of here
was only 13° below freezing; but we are buried in snow !

March 5th.—No alteration in the weather, nor in my affection for my dear niece and nephew and their ten children !
the first is as cold as the latter is warm !

*   *   *   *   *

April 29, 1845.

In his father's library my nephew must have found a
folio volume of H—— (an astronomer and copper engraver),
where, for every hour a distinct picture [of the moon] is
given. In the Phil. Transactions for 1780, p. 507, is the
first paper of William Herschel on the Moon. In 1787;
1792, p. 27 ; 1793, p. 206, measure of mountains, &c.

Twenty-three years ago, when first I came here, I visited
Madame W. (not von) once or twice, saw her observatory
and a telescope, I believe not above 24-inch focal length ;
at that time she amused herself with modelling the heads of
the Roman Emperors : her daughter, then a girl, was a poet,

* The fellow is a fool !

and a portrait of her was exhibited as a Sappho crowned with laurels. . . . .

*   *   *   *   *

The great difficulty of writing begins at last to tell in Miss Herschel's correspondence. One more letter in 1845, is the last of the ample sheets she had been used to fill. The monthly report becomes shorter, more blotted, and betrays extreme feebleness. On the first of October, 1846, she wrote :

MY DEAREST NIECE !—

I must not let the messenger go without a line just to say that I am still in the land of the living, of which, however, I have no other proof than a letter from Baron v. Humboldt, inclosing a Golden Medal from the King of Prussia. I can say no more at present, and the post will not wait, so believe me, my dear niece, yours and my dear nephew's most affectionate aunt,

CAR. HERSCHEL.

The following is the letter referred to from Alexander von Humboldt which accompanied the Gold Medal presented by the King of Prussia on the occasion of her ninety-sixth birthday :—

BERLIN, *Sept.* 25, 1846.

MOST HONOURED LADY AND FRIEND !—

His Majesty the King, in recognition of the valuable services rendered to Astronomy by you, as the fellow-worker of your immortal brother, Sir William Herschel, by discoveries, observations, and laborious calculations, commanded me, before his departure for Silesia, to convey to you, in his name, the large Gold Medal for Science, and to express to you the gratification he felt that, by God's grace, your noble life has been a long succession of years free of pain, and that now in your solitude you continue to enjoy the reflected glory of the all-embracing knowledge, the great labours in both hemispheres, and the profoundly penetrating genius of your illustrious nephew, Sir John Herschel. To be had in remembrance by an intellectual and kind-hearted Prince cannot be a matter of indifference to you. He had wished you to receive this little gratification on your ninety-sixth birthday, and by an unfortunate mistake the date of Caroline Lucretia Herschel's birth has been changed from the 16th of March to the 16th of October, and *I* am the culprit, misled by a misprint in a French history of astronomy. I know I may count upon your indulgence and that of your distinguished family in England. I specially deserve such leniency to-day—the day on which my young friend, Dr. Galle, assistant astronomer in our Observatory (to the triumph of theoretical astronomy be it said), has discovered the transuranian planet indicated by Leverrier as the cause of the perturbations of Uranus.

With the deepest respect,

I am your most obedient, although illegible,

ALEXANDER HUMBOLDT.

Do not trouble yourself to write to the King; I will convey your thanks to him.

Once more a few lines, begun November 1st, and finished December 3rd, were traced, betraying, now only for the first time, the apprehension that they might be the last, in the words—

Miss Beckedorff shall write for me if I do not get better. Loves to *all.*

CAROLINE HERSCHEL.

z

Even this, the last letter of all, is addressed in a large, clear handwriting.   Henceforth "the messenger" carried no more the large familiar sheet which had often been filled at the cost of many days' work and frequent re-writing; but her kind friend, Miss Becke-dorff, wrote a regular monthly report to the anxious friends in England, from which the following most interesting extracts are taken :—

EXTRACTS FROM THE LETTERS OF MISS BECKEDORFF TO
SIR JOHN AND LADY HERSCHEL.

Dec. 1846.

*   *   *   *   *

. . . . She said that whilst she was idling away her time on her couch she had—with her mind's eye—set up a whole solar system in one corner of her room, and given to each newly-discovered star its proper place.   She cried when I told her again of your and Sir John's solicitude about her, &c.

March, 1847.

Her likeness has been taken by two young painters lately. . . .   She was sitting—or rather reclining—for her picture whilst my niece was with her, and the exertion of it made her at first nervous and hysterical, but by degrees she overcame it, and conversed cheerfully.   I am sorry to say the drawing which I saw did not do justice to her intelligent countenance; the features are too strong, not feminine enough, and the expression too fierce; but I hear the picture which I did not see is more like her.

March 31, 1847.

I am commissioned by dear Miss Herschel to send to you and for her dear nephew, with her best love, the accom-

panying print, which I fear will at first sight not satisfy you.
The artist has, I believe, imitated the style of the old German
school of Albert Dürer, resembling more a 'woodcut'
than a print, nor does it justice to her fine old countenance.
Yet it is extremely like in features, expression, and deport-
ment, her eyes having taken the languid expression more
from fatigue occasioned by her *sitting* for the picture, whilst
she is used generally to recline on her sofa, and I see them
very frequently sparkle with all their former animation. . . .
She has, as I predicted, lived to begin her ninety-eighth
year, and she has stood the exertions and excitements of her
birthday even better than could have been expected. I saw
her on the 15th, and again on the 17th; for knowing that
Mrs. Clarke, who, like all General Halkett's family, are full
of kind attentions to her, would act as her aide-de-camp on
the occasion, I felt that it would only be adding to the
number of those who must be kindly spoken to if I had
gone to see her on the 16th. Upon passing the door I just
saw a beautiful and most comfortable velvet armchair, a
cake, and magnificent nosegay carried up to her, and soon
after met the gracious donor, our kind Crown-Princess, with
the Crown-Prince and the Royal child driving to her; they
stayed nearly two hours, Miss Herschel conversing with
them without relaxation, and even singing to them a composition of Sir William's, ' Suppose we sing a Catch.' The
King sent his message by Countess Grote. On the 17th I
found her, more revived than exhausted, in a new gown and
smart cap, which Betty provided; and Betty's own cap was
new trimmed for the occasion, strictly in keeping with the
style of her mistress, and I can but again commend the
judgment and zeal with which she makes her arrangements
for the comfort and appearance of dear Miss Herschel, and
for a fit reception of her high and numerous visitors.
    . . . . I ran over to ask for Miss Herschel's own message

before I seal. I am to "give her best love to her dear
nephew, niece, and the children, and to say that she often
wished to be with them, often felt alone, did not quite like
old age with its weaknesses and infirmities, but that she too
sometimes laughed at the world, liked her meals, and was
satisfied with Betty's services."

. . . . You may rest assured that she is most carefully
attended to, and Betty is not only fully to be depended
upon, but is also extremely judicious, and the only person
who has gained Miss Herschel's entire confidence and
approbation. . . . . I have charged her to come to me
whenever she sees a possibility of doing anything for her
mistress's comfort, and, from the girl's unaffected attach-
ment for her, can quite rely upon her. Dear Miss Herschel
has, indeed, arranged everything beforehand ; and for years
past has reserved a sum to answer all calls in the event of
her death.

June 29, 1847.

. . . . I generally find her dozing, and now always
lying on her sofa ; she requires, however, but a very short
moment to recollect herself, and then enters into a conversation, of which she takes the greater and by far the better
part on herself. It generally carries her back to old times
and events and persons long gone by, sometimes with great
humour, sometimes with regret ; and when she enters upon
subjects of vexation, I have the means of restoring cheerfulness and satisfaction by speaking of her nephew and his
family. She avoids topics of a directly serious and religious
nature—and is indeed so much alone that she has time for
these reflections when by herself.

Dec. 2, 1847.

A few days ago she talked of her childhood, and even
sung me a little ballad she had then learnt.

While her faculties were equal to the appreciation of the gift, she received a copy of Sir John Herschel's great work of *Cape Observations.* The first of the two following letters tells how it was in progress; the next announces its completion; and thus, by a most striking and happy coincidence, she, whose unflagging toil had so greatly contributed to its successful prosecution in the hands of her beloved brother, lived to witness its triumphant termination through the no-less persistent industry and strenuous labour of his son, and her last days were crowned by the possession of the work which brought to its glorious conclusion Sir William Herschel's vast undertaking—THE SURVEY OF THE NEBULOUS HEAVENS.

SIR J. F. W. HERSCHEL TO MISS HERSCHEL.

COLLINGWOOD, *Dec.* 8, 1846.

MY DEAR AUNT,—

Your letter, which arrived this morning, confirms the apprehension which the absence of any news from you during the last month had begun to excite, that you were unwell, and has caused us the liveliest sorrow. How I wish we were near you, that dear M. could be with you and nurse you. But the same kind Providence which has preserved you so long in health will not fail you in sickness. Meanwhile, I pray and entreat you not to decline the attendance of our good Dr. Mühry, or to avail yourself of any comforts that Hanover can afford. We shall look most anxiously for further accounts from Mde. Knippng, or if her family distresses will not allow her (as you say she has lost her mother very lately), from the kind pen of Miss Beckedorff, and I hope they will not wait for the messenger, but write by the post, and that immediately, as soon as this reaches your hands.

Still I trust to see many more letters in your own handwriting, and that the cessation of the very severe weather we have had of late will prove beneficial in restoring your strength, to enable you to face the farther progress of the season, which, if your climate is anything like ours, is always worse in February than at Christmas. . . . .

I am working still hard at my book (of which you will have by this time received the first four hundred pages), but I cannot get on quite so fast as I would, and I greatly fear it will not be out by Christmas.

SIR JOHN HERSCHEL TO MISS HERSCHEL.

July 11, 1847.

MY DEAR AUNT,—

I send to the messenger who will take this, a copy of my " Cape Observations" for you, and I hope it will not be too large for him to take.

You will then have in your hands the completion of my father's work—" The Survey of the Nebulous Heavens."

I hope you will be able to look at the figures (the engravings of the principal nebulæ). As to the letter-press, the Introduction will perhaps interest you, and I daresay Miss Beckedorff or Mde. Knipping will be kind enough to read it to you—a little at a time.

A copy is on its way I presume by this time to His Majesty the King of Hanover, as a testimony of respect to a sovereign who has shown you on many occasions such kind attentions.

Louisa sends you all our news, and the autographs of

Struve and Adams, who, with M. Leverrier, are now at Collingwood.

Adieu, dear aunt,

From your ever-affectionate nephew,

J. F. W. HERSCHEL.

But the time was past when such gifts could be acknowledged with the old enthusiasm, though the faculty to appreciate them had not failed, and we can well imagine how nothing in the power of man to bestow could have given her such pleasure on her death-bed as this last crowning completion of her brother's work.

The Day-book had long ceased. The final entry, on 3rd September, 1845, is "*Astronomischen Nachrichten* * came in." As the letters show, the never-failing birthday festival had been gallantly encountered, and the accustomed offerings of her many friends with their good wishes, always including those of the Royal Family, received in the usual place. But the curtain begins to descend, and the months to go by with only a bulletin to announce that she still lived, and, as the following extract from a letter written by her friend Miss Beckedorff shows, with unabated will and perfectly collected faculties :—

Her decided objection to having her bed placed in a warmer room had brought on a cold and cough, and so firm

---

* The days on which this periodical arrived are always noted in the Day-books.

was her determination to preserve her old customs, and not
to yield to increasing infirmities, that when, upon Dr. M.'s
positive orders, I had a bed made up in her room, before
she came to sit in it one day, it was not till two o'clock in
the night that Betty could persuade her to lie down in it.
Upon going to her the next morning, I had the satisfaction,
however, of finding her perfectly reconciled to the arrangement ; she now felt the comfort of being undisturbed, and
she has kept to her bed ever since.   Her mental and bodily
strength is gradually declining, and although she at times
rallies wonderfully, we can hardly expect that another
month will elapse ere I have to make my sad and last
report. . . . She says that she is without pain ; fever has
left her, and her pulse is regular and good, though weak at
times.   She still turns and even raises herself without
assistance, and at times converses with us. . . . . A few
days ago she was ready for a joke.   When Mrs. Clarke told
her that General Halkett sent his love, and " hoped she
would soon be so well again that he might come and give
her a kiss, as he had done on her birthday," she looked very
archly at her, and said, " Tell the General that I have not
tasted anything since I liked so well."   I have just left her,
and upon my asking her to give me a message for her
nephew, she said, " Tell them that I am good for nothing,"
and went to sleep again. . . . . She is not averse to seeing
visitors.

January 6th.

Four days later the same kind friend had to tell
how peacefully and gently the end came at last.

Jan. 10*th*, 1848.—Your excellent aunt, my kind revered
friend, breathed her last at eleven o'clock last night, the 9th
of January. . . .   She suffered but little, and went to sleep

at last with scarcely a struggle. Up to the last moment she has had the most undeniable proofs of the affection and veneration of her own family and a number of friends, both English and German. Mr. Wilkinson, the English clergyman, has been unremitting in his visits, and so kind and judicious was his manner, that she received them to the last with unfeigned satisfaction. . . . . At four o'clock the guns announced the birth of a young Princess—an event she had anticipated with much interest; and upon her being told of it she opened her eyes for the last time with consciousness.

The following, translated from a letter of Miss Herschel's niece, Mrs. Knipping, to her cousin, Sir J. Herschel, is a most precious fragment, expressing the sentiments of one who for years contributed to lighten the grievous burden of age and growing infirmity by her constant affection and appreciative sympathy. The regret that so little remains from the same pen is enhanced by the fact that no notes, or memorials of any kind, appear to exist by which we might hope to picture to ourselves one whose unconscious self-portraiture makes us crave to see and know and become familiarly acquainted with her, as she was seen and known by others. Comparatively recent as was her death, to the best of our knowledge all have passed away from whose lips we could hope to gather the impressions of personal acquaintance. Excepting from the letters already quoted on the occasion of her nephew's two visits to Hanover, it is not until she lay on her deathbed that we obtain a glimpse of her drawn by any other hand than her own.

January 13, 1848.

. . . . I felt almost a sense of joyful relief at the death of my aunt, in the thought that now the unquiet heart was at rest. All that she had of love to give was concentrated on her beloved brother. At his death she felt herself alone. For after those long years of separation she could not but find us all strange to her, and no one could ever replace his loss. Time did indeed lessen and soften the overpowering weight of her grief, and then she would regret that she had ever left England, and condemned herself to live in a country where nobody cared for astronomy. I shared her regret, but I knew too well that even in England she must have found the same blank. She looked upon progress in science as so much detraction from her brother's fame, and even your investigations would have become a source of estrangement had she been with you. She lived altogether in the past, and she found the present not only strange but annoying. Now, thank God, she has gone where she will find again all that she loved. I shall long feel her loss, for I prized and loved her dearly, and it is to me a most precious recollection that she loved me best of all those here, admitted me to closer intimacy, and allowed me to know something even of her inner life.

All the necessary instructions about her property, her house, her burial, she had written years before; even the sum which she considered sufficient had been carefully set apart for the funeral expenses, and everything, down to the minutest trifle, had been arranged, so that her executor, Sir John Herschel, might have

the least possible trouble. She especially prayed him
not to come should her death occur in the winter; but
the reiterated instructions through the long series of
letters show how keen was her anxiety that whatever
she possessed of value should pass into his hands, and
that no one of her Hanoverian connections, with the
exception of Mrs. Knipping [who, with Miss Beckedorff, was entrusted with her keys], should inter-
meddle. She desired to be laid beside her father and
mother, and an inscription * of her own composition
records how she was her brother's assistant, &c. She
was followed to the grave by many relations and
friends, the Royal carriages forming part of the pro-
cession; the coffin was covered with garlands of
laurel and cypress and palm branches sent by the
Crown Princess from Herrnhausen, and the holy
words spoken over it were uttered in that same
garrison church in which, nearly a century before, she
had been christened, and afterwards confirmed. One
direction she could not put on paper, but she desired
Mrs. Knipping to place in her coffin a lock of her
beloved brother's hair and an old, almost obliterated,
almanack that had been used by her father.

* The inscription is given in the Appendix.

APPENDIX.

THE inventory of the books, pictures, &c., in the sitting-room of No. 376 Braunschweiger Strass, is too characteristic to be omitted. The following is a copy of it :—

Inventory of engravings, all in good black frames, with gilded beads, and glazed :—
My Nephew, J. H.
My Mother.
A drawing of Slough, by J. Herschel.
My Brother, Lithographed.
Forty-foot Telescope.
Medallion of Wm. H., by Flaxman, of 1782.
      Ditto,   by Lochie,   of 1787.
Engraving of Dr. Maskelyne, and
Greenwich Observatory (presented to me by himself).

BOOKS.

Bode's Atlas.
South's Observations on Double and Treble Stars, from Phil. Transactions, Vol. I., 1826.
South's Discordance between the Sun's observed and computed Place. 1826.
On the Elements and Orbit of Halley's Comet, &c., by Lieut. W. S. Stratford, 1837.
Preface to, &c., &c., of a General Astronomical Catalogue, by F. Wollaston, 1789.
J. H.'s Fourth Series of Observations with a twenty-foot Reflector, containing the places of 1236 Double Stars.
Stars in the Southern Hemisphere, observed at Paramatta, in New South Wales, by J. Dunlop, 1828.
Astronom. Nachrichtens, from 1833 to 1839, in 7 vols. (half bound).
Emerson's Treatise of Arithmetic.

Introduction to Sir I. Newton's Philosophy, with an Essay on, &c., by John Ryland, M.A. (Mem.—A Keepsake of General Komerzeusky to me, and now the same to my dear Nephew from his affectionate Aunt, C. H.)

Salmon's Geographical and Astronomical Grammar.

Ferguson's Astronomy.

Watson's Universal Gazetteer.

Quarterly Journal, Vol. XII., 1822.

Quarterly Review, July,1832.

Edinburgh Review, January, 1834.

The Connection of Physical Science, by Mrs. Somerville, 1835.

Third Vol. of Johanna Baillie's Plays. (Mem.—Was given me by Lady H. the day before I left England, to remember my friend, J. B.

John F. Wm. Herschel's Discourses on Nat. Philosophy, which was published in Dr. Lardner's Cabinet, and that on Astronomy, I had handsomely bound and presented them to the Duke of Cambridge, who asked them of me, and would not even wait till I could read them through myself.

Göttinger Anzeigen, 202, 203 Stück, Dec. 14, 1833.

J. Herschel's Papers, from January 12th, 1828, to Nov. 11th, 1833. Bound and directed to the Duke of Cambridge (from C. H.).

Eighteen of Wm. H.'s Papers, collected and bound in one volume, and directed for Hauptman Müller.

Über den Neuentdecken Planeten, by Bode, 1784.

Introduction to English Grammar, by R. South.

1st and 2nd Vols. of Pfaf's Translation of Herschel's Sämptlichen Schriften, 1826 (collected works).

Abominable stuff! What is to be done with them? They are so prettily bound, I cannot take it in my heart to burn them.

Landing place and five back rooms contain nothing but what is necessary for the convenience of my servant and myself; and is mostly bought at the fairs, for a trifling price. (Tables and chairs stained like mahogany, the latter with cane bottoms, at 18d. a-piece, are, after seven years' use, like new.)

Landing-place : A clothes-press, a glass globe, a few chairs.

My Bedroom : A bedstead and bedding, &c., &c. 70 thl. dressing-glass, mahogany frame, plate 22 by 14 inches. (I brought it with me from England.)

A cupboard containing tea things, &c., for company. Urn, tea-board, &c., waiter, two teapots, milk-pot, and slop-bason (black Wedgwood).

666

A few cups and saucers, coffee-pot, two glass plates, one and half dozen bishop glasses, tumblers, cake-basket, &c.

Plate: Ha! ha! ha! ha!

Twelve teaspoons, 1 sugar-tongs, 1 table, 1 dessert, and 1 saltspoon, 4 plated candlesticks, very little used.

The superscription on the last page is as follows:—

It is a pity that I am not at Slough to put the glazed prints in my nephew's study; and many articles of furniture would be so useful in the school-room of my little nephews and nieces. God bless them all!

EPITAPH OF MISS HERSCHEL.

Hier ruhet die irdische Hülle von
CAROLINA HERSCHEL,
Geboren zu Hannover den 16ten Marz, 1750,
Gestorben den 9ten Januar, 1848.

Der Blick der Verklärten war hienieden dem gestirnten Himmel zugewandt, die eigenen Cometen Entdeckungen, und die Theilnahme an den unsterblichen Arbeiten ihres Bruders, Wilhelm Herschel, zeugen davon bis in die späte Nachwelt.

Die Königliche Irländische Akademie zu Dublin und die Königliche Astronomische Gesellschaft in London zählten sie zu ihren Mitgliedern.

In den Alter von 97 Jahren 10 Monathen entschlief sie mit heiterer Ruhe und bei völliger Geisteskraft, ihrem zu einem besseren Leben vorangegangenen Vater Isaac Herschel folgend der ein Lebensalter von 60 Jahren, 2 Monathen, 17 Tagen erreichte und seit den 25ten Marz, 1767, hierneben begraben liegt.

[*Translation.*]

Here rests the earthly exterior of
CAROLINE HERSCHEL,
Born at Hanover, March 16, 1750,
Died January 9, 1848.

The eyes of Her who is glorified were here below turned to the starry Heavens. Her own Discoveries of Comets, and her participation in the immortal Labours of her Brother, William Herschel, bear witness of this to future ages.

The Royal Irish Academy of Dublin, and the Royal Astronomical Society of London enrolled Her name among their Members.

At the age of 97 years 10 months she fell asleep in calm rest, and in the full possession of her faculties, following into a better Life her Father, Isaac Herschel, who lived to the age of 60 years 2 months 17 days, and lies buried not far off, since the 29th of March, 1767.

THE GRAVE OF CAROLINE HERSCHEL.

FROM MISS BECKEDORFF.

Feb. 4, 1850.

" . . . If I have owned my having neglected visiting Sir John's *living* relations, it has not been the same with the churchyard. I have now been confined with cold and fever seven weeks, but one of my last visits was to our lamented friend's grave, which, with the stone and inscription on it, was in perfect order. On the 16th of March I intend to have a bush of white roses planted near it, knowing that my good mother would have paid her that little tribute had she out-lived her revered friend. The white rose she had planted on the grave of Mrs. P. (?) in the same churchyard (the mutual friend of both) continue to blossom every year, and now are a memorial to me and my good mother likewise."

FROM HERR WINNECKE (Assist. Astron. at Pulkowa.)

" Travelling a few days ago through Hanover, I seized the oppor-tunity of visiting Miss Caroline's grave. Pastor Richter, her grand-nephew, took me to it. It is in the churchyard of the ' Garten-gemeinde,' and in a good state of preservation ; a heavy slab lies on it, on which is engraved a long inscription, composed by Miss Caroline herself. At the head is planted a rose-bush, from which I gathered the leaves which I enclose. I venture also to send two ' shadow-outlines' of Miss Caroline, which I had taken from a silhouette in the possession of Frau Dr. Groskopff."

June 26, 1864.

INDEX.

ACADEMY.

Academy, R. Irish, 300.
Astronomical Society of London, 221, 271.
Aubert, Alex., letter from Miss Herschel on discovering her first comet, 66; her third comet, 86.

Baily, F., letter from Miss Herschel, 272–274; letter to her with his "Account of Flamsteed," 281; her answer, 282.
Baldwin, Miss, her marriage, 129; death, 132.
Banks, Sir J., letter from William Herschel on his sister's second comet, 84; from Miss Herschel on her third comet, 85; and her eighth, 94.
Beckedorff, Miss, letters during the latter years of Miss Herschel's life, 338–340, 343–345.
Beckedorff, Mrs., 108.
Blagden, Dr., letter from Miss Herschel about her first comet, 65.
Brewster, Sir David, opinion of Miss Herschel's catalogue of all the star-clusters and Nebulæ, 145, 146.

Cambridge, Duke of, letter to Miss Herschel on the return of her nephew from the Cape, 292.
Cape of Good Hope—Sir John Herschel leaves the Cape, 292.
Collingwood, the seat of the Herschel family, 320.
Comets, Miss Herschel's first, 64; second, 80; third, 85; fifth, sixth, 93; eighth, 94.
Cumberland, Duke of, proclaimed king of Hanover, 290.

Dessau, Princess of Anhalt, letter to Miss Herschel, 267.

[HERSCHEL.

Earthquake at Lisbon, sensation produced in Hanover, 6.
Encke, Prof., letter to Miss Herschel, 248.
Englefield, Sir H., letter from W. Herschel on his sister's second comet, 83.
Epitaph on Miss Herschel, 351.
Etna, Mount, ascent by Sir John Herschel, 173.

Flamsteed's Catalogue, calculations for, 60.
Forty-foot telescope, 76, 308, 309, 310.

Gauss, Hofrath, letter from Miss Herschel, with her index to Flamsteed's Observations, 191; his answer, 195.
George III. visits the Slough Observatory, 104; anecdote of, and the Archb. of Canterbury, 309.
Georgian Satellites, the, 74, 305, 316.
Georgium Sidus, the, discovered, 39.
Gloucester, Princess Sophia of, visit to the telescope, 128.

Halley's Comet, 283.
Herschel, Alex., assists his brother William, 36, 53, 109, 111, 115, 122; returns to Hanover, 125; death, 132; notice of, 132.
Herschel, Caroline Lucretia, early recollections, 1–28; affection for her brother William, 9; at the Garrison school, 11; her father's careful training, 13; typhus fever, 15; confirmation, 17; learns dress-making, 21; accompanies William to England, 26–28; life in Bath, 29–50; *Heimwehe*, 33; visit to

354   *Index.*

HERSCHEL.

Mrs. Colebrook, 34 ; musical rehearsals, 36; reputation as a singer,
40 ; assists her brother, 42 ; life at
Datchet, 50 ; accidents, 55 ; Clay
Hall, 57 ; Slough, 58 ; Flamsteed's
Catalogue, 60, 61 ; her *sweeps*, 64,
146–148 ; first comet, 64 ; salary
of 50*l*. as her brother's assistant,
75 ; her eight comets, 80–94 ; lives
by herself, 95 ; Index to Flamsteed's Observations, 96 ; extracts
from diary, 98–132 ; at Bath, 105 ;
at Slough, 107 ; removes to Chalvy,
108 ; resides at Upton, 109 ; returns to Hanover on the death of
her brother, 133 ; *Recollections*,
133–140 ; her works, 145 ; bitter
disappointment in her brother
Dietrich's family, 149 ; letters,
152 ; Catalogue of the Nebulæ,
181 ; her will, 200 ; presentation
of the Gold Medal of the Royal
Astronomical Society, 221 ; her
portrait, 237, 338 ; Paganini, 247 ;
her nephew's visit, 254 ; anecdotes
of his boyish amusements, 259 ;
Hon. Member of the Royal Astronomical Society, 271 ; letter from
Mrs. Somerville, 274 ; illumination in honour of the Duke of
Cumberland being proclaimed king
of Hanover, 290 ; visit of her
nephew and his son, 293–295 ; Hon.
Member of the R. Irish Academy,
300 ; extracts from day-book, 303–
307 ; anecdotes of the forty-foot
telescope, 308, 309 ; describes
Christmas in Germany, 313 ; her
92nd birthday, 318 ; begins the history of the Herschels, 324 ; her
93rd birthday, 330 ; the first
railway between Hanover and
Braunschweig, 334 ; presented with
a gold medal by the king of
Prussia, 336 ; her last letter, 337 ;
enters her 98th year, 339 ; her
death, 344 ; funeral, 347 ; epitaph,
351 ; her grave, 352.
Herschel, Sir John, first mention of,
104 ; at Cambridge, 117 ; senior
wrangler, 120 ; member of the University of Göttingen, 125 ; ascends
Mount Etna, 172 ; at Munich, 175 ;

LA LANDE.

visits his aunt, 177, 293 ; Secretary
to the Royal Society, 181 ; at
Montpelier, 201 ; catalogue of
double stars, 213 ; his marriage,
236 ; describes his aunt, 254 ; anecdotes of his boyhood, 259 ; letters
from the Cape, 263 ; sweeping,
266 ; the Milky Way, 270 ; Halley's
comet, 283 ; spots on the sun, 286,
287 ; Saturn's sixth satellite, 288,
289 ; returns to England, 292 ;
created a baronet, 305 ; on the
Orionis star, 316 ; eclipse of the
sun in 1842, 327 ; his chrysotype
pictures, 327 ; translation of Schiller's "Walk," 328, 329 ; acknow-
ledges his aunt's history, 333.
Herschel, Lady, letters from Miss
Herschel, 152 *et seq.* ; her death,
252.
Herschel, Sir William, early display
of talents, 3 ; proficiency in music,
7 ; accompanies his regiment to
England, 8 ; resides at Bath, 21 ;
fetches his sister Caroline, 26 ; his
musical compositions, 36 ; erection
of the twenty-foot telescope, 37 ;
discovers the Georgium Sidus, 39 ;
casting of the great mirror, 43 ;
goes to London and is introduced
to the King, 45 ; Royal Astronomer,
50 ; limited salary, 50 ; removes
to Datchet, 50 ; to Clay Hall,
57 ; to Slough, 58 ; the Georgian
Satellites, 74 ; marriage, 78 ; observations on his sister's comet,
84, 85 ; his failing health, 124 ;
sits for his portrait, 129 ; death,
133.
Hesse, Princess of, letter to Miss
Herschel, 267.
Humboldt, Alex. von, letter to Miss
Herschel, with the Gold Medal for
Science from the king of Prussia,
336, 337.

Knipping, Mme., extract from letter
upon Miss Herschel's death, 346.

Lind, James, 100.
La Lande, J. de, letter to Miss Herschel, 89 ; her answer, 91.

MORGAN.

Morgan, A. de, letter from Miss Herschel on being elected Hon. Member of the R. A. Society, 271.
Mars, observations on, 53.
Maskelyne, Rev. Dr., letter from Miss Herschel, on discovering her second comet, 80 ; on the Index to Flamsteed's Observations, 96.

Nebulæ, the, 196–198.
Nebulæ, the Cape, and double stars, 328.

Ole Bull, the violinist, 306.
Orange, Prince of, at Slough, 99.
Orionis, α, a variable and periodical star, 316.

Piazzi, Abbé, at Slough, 55 ; at Catania, 173.
Pigott, Ed., letter to Miss Herschel on the Flamsteed Catalogue, 101.

Railway, first, between Hanover and Braunschweig, 334.
Ross, Capt., his return with the South Polar Expedition, 333, 334.

Schiller's " Walk," translated by Sir J. Herschel, 328, 329.

ZODIACAL.

Schumacher, Prof., letter from Miss Herschel, 260.
Scorpio, 258, 266.
Seyffer, Prof., letter to Miss Herschel, 92.
Somerville, Mrs., letter to Miss Herschel, with her " Connexion of the Physical Sciences," 274.
South, J., his 400 stars, 194 ; his address to the Astronomical Society on presenting the hon. medal to Miss Herschel, 222–227.
Stewart, P., letter from Miss Herschel, 277.
Sun, spots on the, 286, 287.
" Survey of the Nebulous Heavens," the conclusion of Sir W. Herschel's vast undertaking, 341.
Sweepings for comets, 146–148.

Telescope, the forty-foot, anecdotes of, 308, 309 ; its final preservation, 310.

Watson, Sir W., first acquaintance with W. Herschel, 42.
Wilson, Alex., notice of, 99.

Zodiacal light, the, 331.

THE END.

BRADBURY, AGNEW, & CO., PRINTERS, WHITEFRIARS.

Printed in the United States
By Bookmasters